GRUNDIG Satellit

All Models in Word and Pic

Second Edition

By Thomas Baier

Grundig Satellit, All Models in Word and Pic
By Thomas Baier

Second Edition Revised, 2011

Front Cover: Grundig AG, Fürth / Bavaria © 1989
Title Page: Grundig AG, Fürth / Bavaria © 1989

® Grundig Satellit, Yacht Boy, Concert Boy, Melody Boy and Ocean Boy are registered Trademarks by Grundig AG, Fürth / Bavaria, Germany

Published by Universal Radio Research
ISBN 978-1-882123-01-8 1-882123-01-8

#2591

Supported by:

Grundig AG, Fürth / Bavaria; Germany

Short-wave listeners club Saar,
(SWLCS), Merchweiler / Saar; Germany

Mr. Mike Petrozello, Ballston Lake / NY; USA
(language support)

Grundig Transistor Radio Forum (GTR)
Adriano, Chuck, Frank, Harvey, Jim, John, Jose, Jules, Ian, Michel, Mike, Ray, Robert, Steve, Tim, Thomas, Tom & Paul

This book is available in German language too.
Also the book "40 Years Grundig Ocean Boy"

Special thanks to my wife for her great patience, and my little son. His dad hasn't got much time to play with him, so he had to play with dad's toys...

Preface

In the early sixties Grundig planned to put a portable radio especially made for short wave reception on the market. The wish of guest worker families of that time to keep in contact with their home countries brought Grundig´s idea ahead. To guarantee a good short wave reception the circuit principle called "Superheterodyne" or simply "Superhet" was used. Accordingly, the radios of that time were called "Travelsuper". The new product was supposed to have a modern and appropriate name. The world was astonished by what the Soviet Union and later the USA were shooting up into space, while in Bavaria in autumn 1964 the first satellite reception facility began to work. So the slogan was found. It was Max Grundig himself who decided to call the new series of radios "Satellit". For a further differentiation within the series he chose the modern term of "Transistor": If the series was to be sold, it seemed more beneficial to call the radios Satellit Transistor 5000 for example than simply calling it Satellit 205. According to that, the first series were also sold in foreign countries as Satellit Transistor. The three figures used in the name of the devices produced for the German market reveal the year of construction. In the fifties, portable radios made by Grundig had quite melodious names with some figures behind them to indicate the year. An example for that is Melody Boy 58. In the sixties, the second decade of production, a 2 was placed in front of the year, a 0 was put in when necessary. Some examples are the Ocean Boy 20**4** = 196**4**, Satellit 20**5** = 196**5**, etc. If we take a closer look at the Grundig Satellit series, we´ll soon become aware of Grundig´s strategy, that is the idea of the double models that was constantly realised. After a short time a new model was replaced by an improved one. This idea proved to be a success. Thus the following pairs emerged: 208 / 210, 2000 / 2100, 3000 / 3400, 1400 / 2400, 300 / 400, 600 / 650 and the devices 500 / 700. While the first type, the Satellit 205, had borrowed much aspects from the Ocean Boy 204, the Satellit 4000 was rather a "descendant" of the Concert Boy 210 and the Satellit 1000 a prudent mixture of well-tried aspects and new ones. There was one thing all the Satellit models that were produced and sold had in common, that is the aerial on the right side.
The author, September 2001

About 2nd edition

In the last 12 month I got much advanced information about Grundig Satellit, i.e. factors of production and technical background. Even I got more radios to provide never seen pics; i.e. from Satellit 900. To ensure a better quality of the photos, a high resolution white paper has been used. All stated prices are based by many "deals" in the internet. Sure, those radios are much cheaper in Germany, but even here the prices are rising. Some friends asked me to print the book in colour. Due to extremely printing costs, price would jump up to $50 USD!! Use my url to see colored pics please. The First Edition, published in the style of the sixties, was financed by my own means.

The expert

Gerhard Weber has been working for Grundig since 1962. He was involved in the development of all the satellites. Till the late seventies the whole series was produced in the production hall A in the Kurgartenstraße 37 in Fürth / Germany, while the development department was working three floors upwards in the so-called Lab 1. In 1983, Lab 5 took over this work, because Lab 1 was overstrained. As a result, the Satellit 300 and 4000 appeared. The last model that produced in Germany was the Satellit 3400, that in the end was produced in Landau / Isar. In the nineties, the development labs were transformed into the Car Radio Development Ltd., where Mr. Weber is working as head of the measuring department. In mid-may 2000, this site moved to Nürnberg / Langwasser. Mr. Weber is in close contact to a few SW listener clubs. In private he often listens to shortwave radio. And he surely hasn´t any need of devices to do so!

Satellit	**205 (Transistor 5000)**
Build	End 1964 to end 1966
Dimensions	41 x 25 x 12 cm
Weight	6,2 kg
Body	Wood, covered by grey vinyl
Ranges	FM, MW, LW, SW
SW-Bands	10, 6 of it at the Drum
SW-Range	160 to 10 mtr single conversion
Drum-Tuner	52 to 16,5 mtr
Bandspread	no
Band Width	one
Display	Analogous and Drum
Output-Power (Sine-power)	2 Watt Battery-operated 2 Watt Main-operated
Aerial (Telescope)	12 Elements
Powerpack	TN12, inserted
SSB	separately
Clock	no
Timer	no
Memory space	none
Dial illumination (Battery operated)	by pushing the volume control
Connection-Sockets	2 DIN Sockets (Line in/out) Earphone 9 V external Power Pack External Aerial
Accessories	Carrying Case Earphone 203 A
New price	DM 598,-

In autumn 1964 Grundig presented a new portable receiver, based on the Ocean Boy 204. Travel super with short-wave already existed. The novelty of this device, with the international name Transistor 5000, was the use of a drum tuner for some of the short-wave-frequencies. Six short-wave tuners were positioned on a drum, similar to a revolver. This way, the whole scale width for the respective selected short-wave band was available. That resulted in a substantially finer and more exact tuning. At that time, the device was state-of-art. Continuous receipt of the short-wave from the 10 mtr band to the long-wave, a fine tuning within that part of the short-wave that was not located on the drum, so called "Kurzwellenlupe". The drum tuner is located on the upper left hand side, arranged horizontally above the loudspeaker The Satellite 205 resembles very much the Grundig Ocean-Boy series. The model Ocean-Boy 207 looked very much like the Satellit 205, except it didn't have a drum tuner and was a little bit smaller. Both devices has push buttons to select the bands and the external antenna in a vertical row at the right. Another feature: When operated on batteries, the lightning of the scale was turned on simply by pressing the volume control. This radio has 2 loudspeaker. A tweeter is hooked up in the main loudspeaker, tweeter is useful for FM. To access the amplifier two German DIN sockets are in use: Line-in and Line-out. In summer 1965 Satellit 205 Amatuer was presented. It's actually the same device, but in addition it receives HAM frequencies and a build-in SSB section. This was firmly inserted on the back side behind the rear hatch. Therefore, the handling was a little bit complicated. Before cleaning, it's recommended to separate the body from the electronic part, which is simple to do. However, the imitated leather strap is used as hinge of the back part. So be cautious when taking it off. A well preserved Sat 205 / TR 5000 is difficult to find on exhibitions or fairs. The international model is called Transistor 5000, the national model Satellit 205. Satellit 205 / TR 5000 has been modified lightly in fall 1965, for example the back hatch is hold by two screws. In excellent condition it's worth 200 US Dollar and more. Generally the rarest Satellit is considered to be the 205 / 5000 Amateur. $500 USD is it worth, with case & papers even more!

Satellit 205
Similarity to the Ocean Boy series.
A novelty : The Drum tuner. The rear hatch contains a device to plug in the single-side-band module.

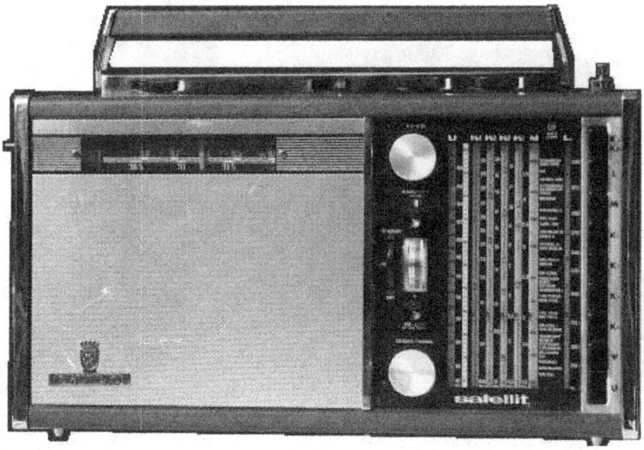

Satellit 205
in the right: The main scale for FM and AM
in the left: The drum tuner

Photo below: The power pack TN 12

Satellit 205
right: Satellit 205; 1965 (Transistor 5000)
left: Ocean Boy 207; 1967 (Transistor 3000)

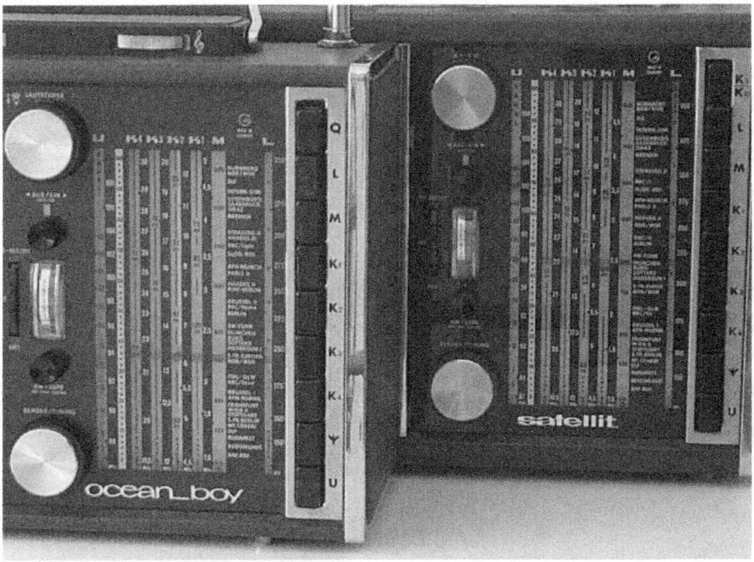

Satellit 205 A (Amateur version)
Top picture: Operation elements for the SSB unit (One screw back hatch model!)
Bottom picture: Built-in SSB unit.

Satellit	**208 (Transistor 6000)**
Build	1967 to end 1968
Dimensions	44 x 26 x 13 cm
Weight	6,2 kg
Body	Wood, covered by vinyl or wooden-like
Ranges	FM, MW, LW, SW
SW-Bands	9, 8 of it at the Drum
SW-Range	160 to 10 mtr
Drum-Tuner	52 to 16,5 mtr double conversion
Bandspread	yes, by button
Band Width	two
Display	Analogous and Drum
Output-Power (Sine-power)	2 Watt Battery-operated 2,5 Watt Main-operated
Aerial (Telescope)	13 Elements
Powerpack	TN12, inserted
SSB	separately
Clock	no
Timer	no
Memory space	none
Dial illumination (Battery operated)	by pushing the red button
Connection-Sockets	1 DIN Socket (Line in/out) Earphone 9 V external Power Pack External Aerial SSB unit
Accessories	Carrying Case Earphone 203 A
New price	DM 698,-

This „super radio" didn't have much in common with its predecessor. A completely new device called „Transistor 6000" was put on the market. The drum tuner was installed vertically right behind the main scale and therefore invisible from the outside. The bands of 49, 41, 31, 25, 19, 16, 13 and 10 metres were integrated as S2 to S9 on the drum. The S1 range (the 60 to 160 metres band) had to be tuned through the main scale. One oscillator frequency (single conversion) was enough to do that. Fine tuning for the S1 range was already known and also integrated. The new feature was the band spread of the S2 to S9 ranges. While pressing the band spread key it was possible to get out the radio and amateur radio bands of all the 8 ranges and to spread them through the whole scale of the drum tuner. This guaranteed a very fine tuning. The SW display, which was also a drum, was on the left of the main dial, also arranged vertically and parallel to the drum tuner. This drum display was driven by a small tooth belt which was in between the drum tuner and the drum display. Actually, this small plastic toothed belt is the most vulnerable part of the radio. It's important that the drum display and the drum tuner always move synchronously and that the display can't slip through. If you want to buy a used set, it's advisable to try to switch this part quickly several times. The keys to control the wave ranges were placed on the upper side of the device from this series on. On the very left side there was a small red key for momentary dial illumination during battery use. The three output controls for volume, bass and treble stayed in the same place. So it was easy to exchange the set's back which, from then on, was fixed with two small screwed hinges. On the back of the device behind a small plastic hatch there was a rectangular part to plug in the supplementary SSB unit. The international model is called Transistor 6000, the national model just Satellit 208. Only national models had the European radio stations marked on the main scale. Otherwise only figures were shown. The two German normed sockets (DIN for Line in and Line out) were combined to only one. The price of a new S 208 was at DM 698 at that time. Today used sets without defects are rare and they are sold for up to $400 USD.

Satellit 208
Wooden-like

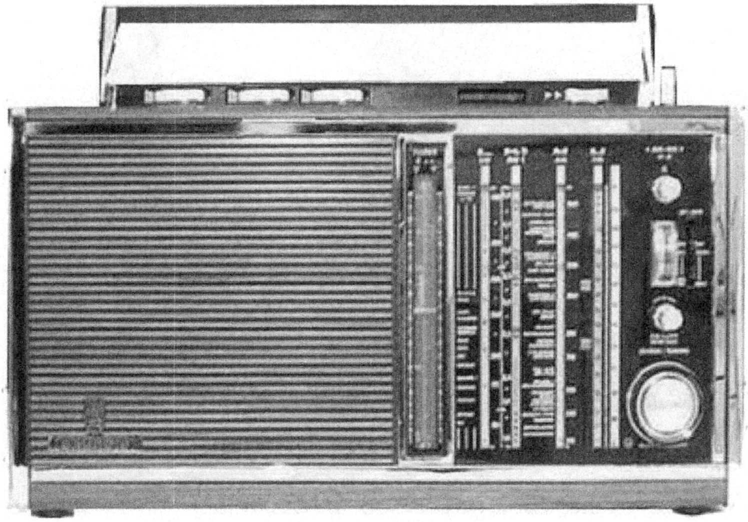

Photo below: black vinyl model with rainbow grill

Satellit 208
Owner's manual in 1967

Photo below: Connection Sockets

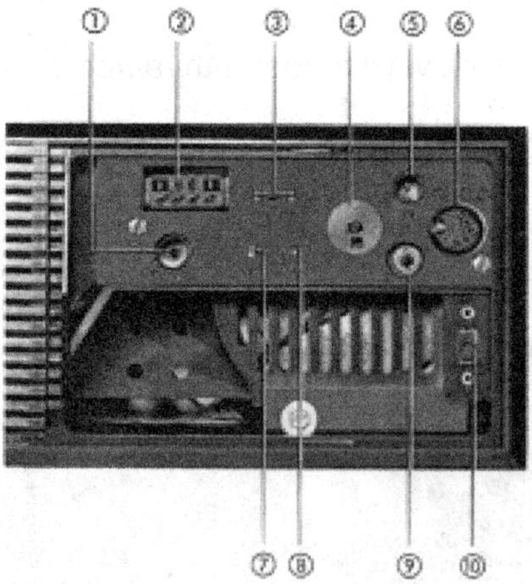

Rear Connection Socket
(see illustration above)

① Car-antenna
② SSB-Kit
③ FM-dipole
④ Satellite speaker (approx. 4 Ω)
⑤ Power pack or car battery
⑥ Record-player or tape recorder
⑦ Antenna for LW, AM and SW
⑧ Ground
⑨ Earphone (GRUNDIG type 203 A)
⑩ Battery/Line Switch

Satellit 208

Left: Grey-blue body; Right: black covered series

Photo below: Closed slide hatch

Satellit 208 / 210 / 1000
The upper turnable switch is to select the band on drum-tuner. The lower one pulls out and for tuning the selected band on the drum tuner. Satellit 208 doesn't have the indentation, see right pic.

Satellit 208 to 2100:

This yoke was fixed by a srew into the radio. The SSB-unit is attached by sliding it over the protruding end of the yoke.

Satellit	210 (Transistor 6001)
Build	1969 to end 1971
Dimensions	44 x 26 x 13 cm
Weight	6,2 kg
Body	Wood, covered by black vinyl or wooden-like
Ranges	FM, MW, LW, SW
SW-Bands	9, 8 of it at the Drum
SW-Range	160 to 10 mtr
Drum-Tuner	52 to 16,5 mtr double conversion
Bandspread	yes, by button
Band Width	two
Display	Analogous and Drum
Output-Power (Sine-power)	2 Watt Battery-operated 2,5 Watt Main-operated
Aerial (Telescope)	13 Elements
Powerpack	TN12 a, inserted
SSB	separately
Clock	no
Timer	no
Memory space	none
Dial illumination battery operated	by pushing the red button
Connection-Sockets	1 DIN Socket (Line in/out) Earphone 9 V external Power Pack External Aerial SSB-unit
Accessories	Carrying Case Earphone 203 A
New price	DM 698,-

Basically, this device was the same as the S 208, but it showed numerous little improvements and modifications. It were just these little changes that made this type such a success. It was easier to connect the headphones at the front of the device. A scale reaching from –5 to +5 was installed for the fine tuning. During battery supply the dial illumination could be switched on constantly by pressing the small red key and turning it right. Another characteristic of the S 210 was that it had four turn controls on the upper side. The fourth and added one served to tune an external aerial. One more feature; the better band spread. On the satellit 208 the band scales are compressed at the top of the scale and spread out at the bottom. The satellit 210 has the scales more evenly spaced, and the Satellit 210 Amateur has the most advanced scales for that time. The radios were available in two different designs, either with a black imitation leather body and a metallic loudspeaker panel with this special glow or an all wooden body. The special panel glow refracts the light and produces a multiple arc (rainbow grill). But if the owner treated the panel with a detergent – hopefully only on the outside – the effect vanished and the panel stayed tarnished. So this indicates clearly that the device has been "treated", what lowers the value of the radio considerably, because the panel can only be replaced by taking the wooden body to pieces. A damaged aerial lowers the price again. An original aerial costs up to $50 USD. The TN 12a transistor power pack were inserted and a power control switcher was integrated in the cord. If your radio sounds "woofy" – check the fuse in the TN 12a. Used 210's in good condition cost US $ 200. Obviously, a collector´s item is more expensive. The radio was sold international as "Transistor 6001" and an amateur radio version was available with amateur radio bands on the drum instead of SW bands and a supplementary SSB unit. The scale only indicated figures (same as Transistor 6000), the band spread key was used to switch the band width on AM. The 210 Amateur has an advanced spreaded scale, even better than regular 210. The Satellit 210 (not the amateur version) was quite common. Black / Silver models are more wanted as wooden-like cabinets.

Satellit 210
Above: wooden-like, below: black vinyl covered

Satellit 210

On the left: The visible drum
On the right: the white drum tuner
Below: a toothbelt between both drums

Photo below: Shortwave guide (1969)

Satellit 210
Photo above: Earphone socket now on the front
Photo below: Satellit 210 Radio Amateur

Satellit 208 (left), 210 (middle), 210 Amateur (right)
49 mtr Band

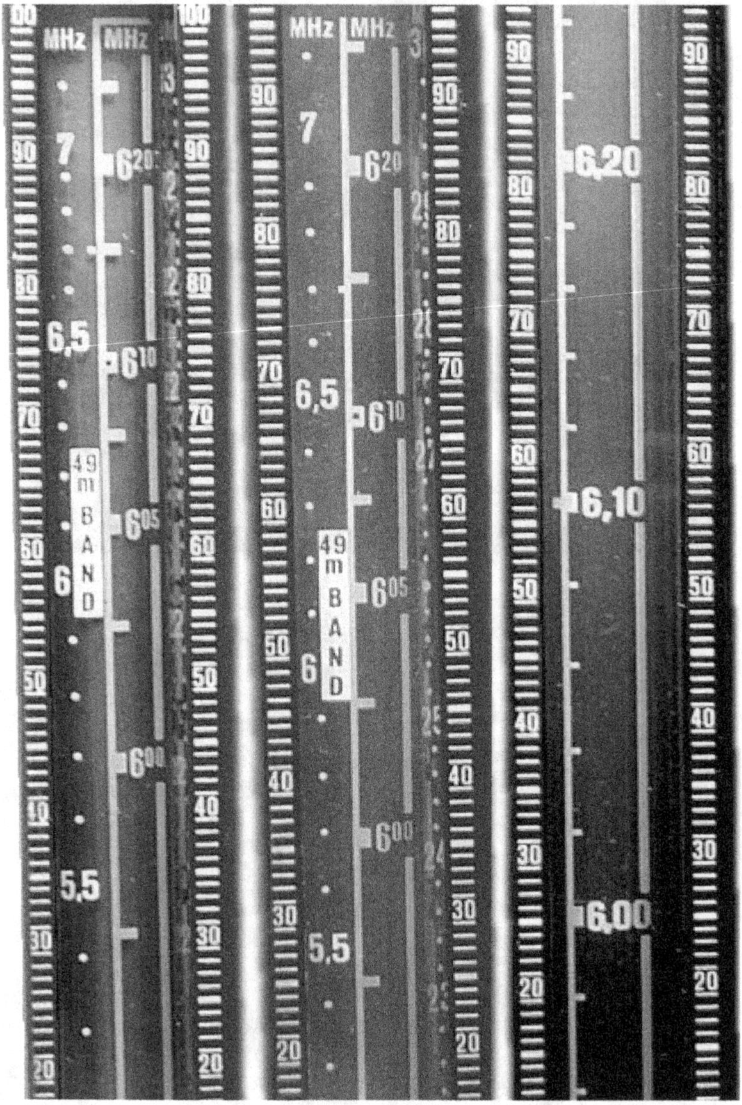

Satellit	**1000 (Transistor 6002)**
Build	1972
Dimensions	47 x 27 x 12 cm
Weight	6,5 kg
Body	plastic (black)
Ranges	FM, MW, LW, SW
SW-Bands	9, 8 of it at the Drum
SW-Range	160 to 10 mtr
Drum-Tuner	52 to 16,5 mtr
double conversion	
Bandspread	yes, by button
Band Width	two
Display	Analogous and Drum
Output-Power	2,5 Watt Battery-operated
(Sine-power)	4 Watt Main-operated
Aerial (Telescope)	13 Elements
Powerpack	build-in
SSB	separately
Clock	no
Timer	no
Memory space	none
Dial illumination	by pushing the red
(Battery operated)	button
Connection-Sockets	1 DIN Socket (Line in/out)
	Earphone
	9 V external Power Pack
	External Aerial
	SSB-unit
Accessories	Carrying Case
	Earphone 203 A
	SSB-unit
New price	DM 798,-

The satellites had already been on the market for almost seven years. In the early seventies the design of the devices was changed. The Satellit 1000 was a stage in between. The metallic loudspeaker panel, the controls on the right side of the radio and the vertical scales including the drum tuners were taken over from previous models. What was new and corresponding to that time were the slide controls to regulate volume and tone. But these slide controls, that actually had been introduced against the will of the development department, proved to be unpractical and were therefore replaced again by turn controls when the S 2100 was released. The round keys on the upper side were also replaced, namely by seven rectangular and bigger ones. For the first time the body was made of plastic. The power switch was transferred to the left part of the device's bottom. The aerial tuner was placed exactly where the on / off turn switch had been before. The case size of the Satellit 1000 is similar to the models 2000 and 2100. For this type aluminium instead of chrome was used to plate the side edges. For the first time the name "Satellit 1000" stood on the outside of the device, before the name had been of internal use only. It was written on the front, on the upper side of the main scale. A few improvements and changes were made beneath the body, e.g. the power pack was now fixed and the output unit was improved, too. So the output power even reached 7watt. For the first time the Dryfit 476 accumulator could function without a special power pack. The power pack could be charged automatically. The system started working at a rated voltage of 7.4v and reduced the charging power at 9.2v. It was possible to charge the accumulator by power supply or by external power and it took about 15 hours to charge an empty one. The radio was also available in professional look - with side parts made of mat plastic and a black loudspeaker panel - or with a wooden design. Today it´s difficult to find a Satellit 1000 because not many of this type have been produced. So if the seller knows the value of the device and the radio is in good condition, then it can cost up to $500 USD.

Satellit 1000
First Satellit with a plastic made body!

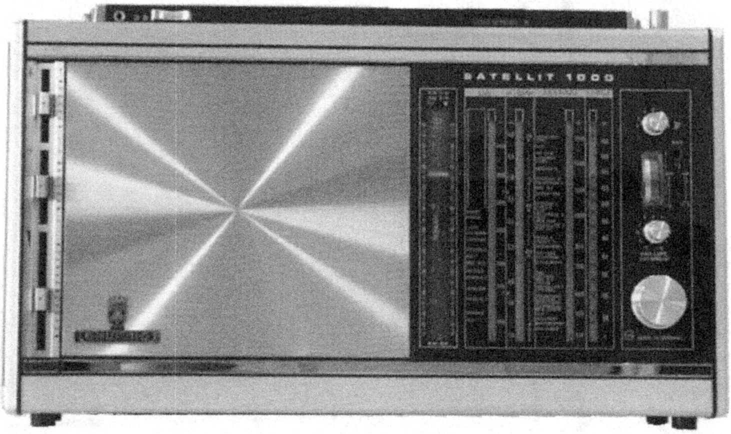

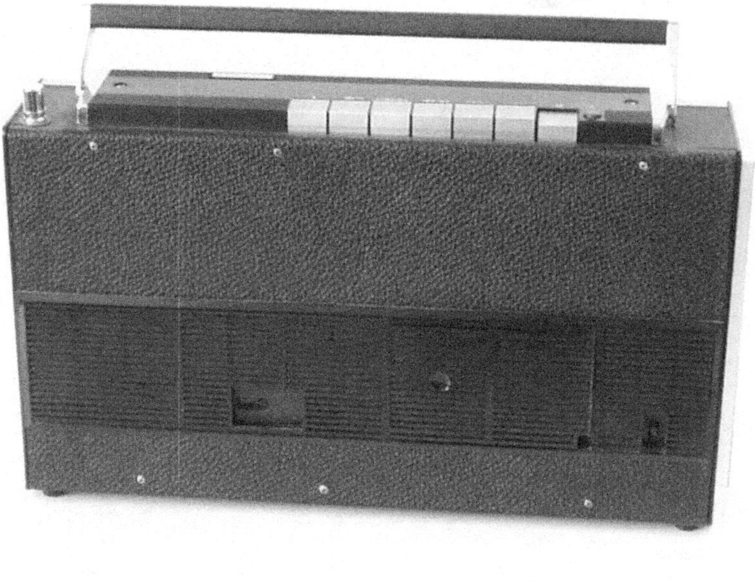

Satellit 1000
Below is shown the volume contol and main power switch.
The lower photo on this page shows the wooden-like.

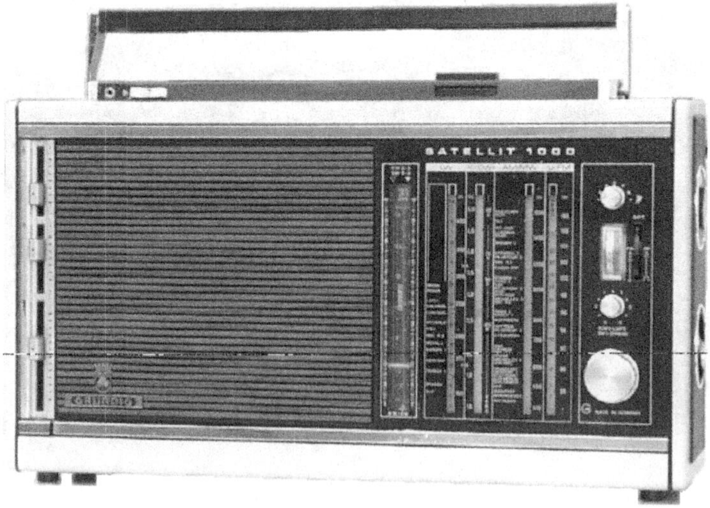

Satellit 1000

The national model (Satellit 1000) is shown above and the international model (TR 6002) is shown below.

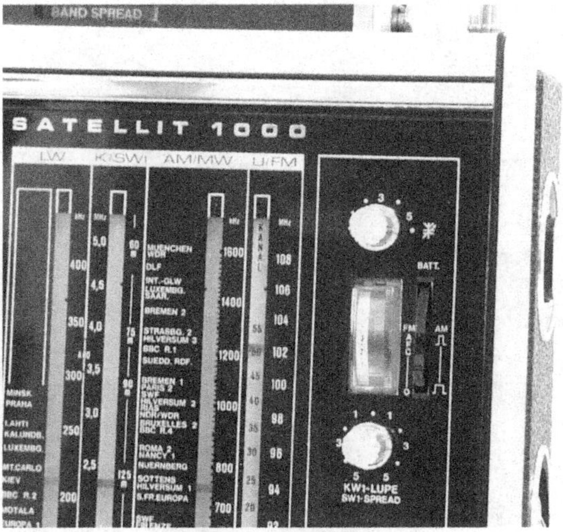

Satellit 1000
Black

Satellit	**2000**
Build	1973 to end 1975
Dimensions	46 x 26 x 12,5 cm
Weight	6,3 kg
Body	plastic
Ranges	FM, MW, LW, SW
SW-Bands	10, 8 of it at the Drum
SW-Range	160 to 10 mtr
Drum-Tuner	52 to 16,5 mtr
double conversion	
Bandspread	yes
Band With	two
Display	Analogous and Drum
Output-Power	2,5 Watt Battery-operated
(Sine-power)	4 Watt Main-operated
Aerial (Telescope)	13 Elements
Powerpack	build in
SSB	separately
Clock	no
Timer	no
Memory space	none
Dial illumination	by pushing
(Battery operated)	lamp-control-button
Connection-Sockets	1 DIN Socket (Line in/out)
	Earphone
	9 V external Power Pack
	External Aerial
	SSB-unit
Accessories	Carrying Case
	Earphone 203 A
	SSB-unit
New price	DM 798,-

This device was the version that followed the S 1000 with three separate dials lying one above another and for the first time these dials were arranged horizontally. The different dials were illuminated according to the range the set was tuned to. The band spread of the SW unit was now fully developed. So the different ranges were less overlapping than they had been in the previous models. A much higher accuracy could be achieved because of the linear dial spread. Actually this dial arrangement made it much easier to work with the dials than it was the case with previous models. It may also be that it was easier to handle because of the simultaneous movement of the tuning control and the dial display moving into the same direction. The SW range 1 was splitted up into two ranges (S1 of 1,6 – 3,5 megacycles per second and S2 of 3,3 – 5,2 megacycles per sec.). The ranges S3 to S10 were controlled through the drum tuner and it was possible to spread them once more. Switching over to another band weight was possible, too. The selectivity on AM was intensified by applying a ceramic filter. The previously used press keys were replaced by a big push-button panel on the upper side of the device, according to the prevailing taste. The small slide hatch on the back that had to be pushed open was replaced by a modern hinged lid. Grundig produced this radio with three different body designs, what is quite remarkable. Since the end of 1972 the S 2000 was available. The loudspeaker panel of this first edition had a polished metal lattice, similar to the previous types. During 1973 this lattice with its special glow was replaced by a plastic one with many small circular holes in it. From 1975 on all the aluminium panels of the body of this type´s last series disappeared. The result of this black version was then sold under the label of "professional look". Once had to pay DM 798 for a new device. Nowadays, satellites of this series are still very common. About $150 USD should be taken into account for a radio that is in good condition, $250 USD for a mint 2000.

Satellit 2000
Above the very first edition, below the second model

Satellit 2000
Last edition, called „Prfi-Look"

Photo below:
The SSB unit

Satellit 2000
Now modern switches like a „ Piano „

positioned as followed

- Shift key for external aerial
- FM (scale below)
- Drum tuner (SW 3 to SW 10) scale in the middle
- SW 2 (scale above)
- SW 1 (scale above)
- Mediumwave (scale above)
- Longwave (scale above)
- Tape / Record player
- Shift-key for tweeter
- Dial illumination when battery operated
- Main AC Switch (On / off)

Satellit 2000
Power pack built-in. The AC cord and plug could inserted to the AC cord box. the closed lid takes care an uniform rear.

The „proper" aerial for Grundig Satellit had 13 elements, and was marked between the 7th and 8th element for FM operation. Except: Satellit 205. He had a 12 element aerial but the same length, the top has an inverted cone shape. Satellit 1400-2400-600-650-800 has the same aerial as shown here. Just the top knob (cylinder) has the half high. Satellit 300-400-500-700-900-4000 doesn't have this type of aerial. They had 8 (sat 300/4000), 11 (400/500/700) respectively 10 (900) elements.

Satellit 2100

Build	1976 to end 1979
Dimensions	46 x 26 x 12,5 cm
Weight	6,3 kg
Body	plastic
Ranges	FM, MW, LW, SW
SW-Bands	10, 8 of it at the Drum
SW-Range	160 to 10 mtr
Drum-Tuner	52 to 16,5 mtr
double conversion	
Bandspread	yes
Band Width	two
Display	Analogous and Drum
Output-Power (Sine-power)	2,5 Watt Battery-operated 4 Watt Main-operated
Aerial (Telescope)	13 Elements
Powerpack	build in
SSB	separately
Clock	no
Timer	no
Memory space	none
Dial illumination (Battery operated)	by pushing the lamp-control-button
Connection-Sockets	1 DIN Socket (Line in/out) Earphone 9 V external Power Pack External Aerial SSB-unit
Accessories	Carrying Case Earphone 203 A SSB-unit
New price	DM 898,-

Not many parts were changed. The amplifier unit was equipped with silicon transistors that emitted less noise. Besides, it got an input stage which reduced the noise output at low volume levels and while using headphones. The output unit was then again operated by turning the controls for volume, bass and treble. This was Grundig´s reaction on the wishes and needs of their customers! Grundig spoke of an output power of 2,5 watt during battery supply and 4 watt during power supply (sine output power). Since 1978 the device's design changed a bit. From then on the loudspeaker panel was known as the "waffle" (see also the chapter about the S 3000) and the knobs were silver. Just as it had been the case with all the previous types, the SSB unit was also a supplement to this model. The last type that was equipped with this supplementary unit was the S 2100. The German standard socket in combination with the TA/TB key made it possible to only use external supply, e.g. for record playing. Just as the previous model, it could be operated by using the Dryfit 476 accumulator because the integrated power pack could be charged automatically. The loudspeaker of this radio was also provided with a supplementary treble loudspeaker, that could be turned off. During AM drive the treble loudspeaker wasn't used anyway. The SW / FM aerial was already known because it had been used for the previous models, except for the S 205. The first seven elements were needed for FM reception. One had to draw out the last six elements for SW reception. This satellite was the last type functioning completely by analogy. No chips, no digital read-out, no counter. At the end this radio cost DM 898. It was a cheap alternative to the Satellit 3000. The very last model is rare today, because for that time, most listeners bought the Satellit 3000 with frequency counter. From today´s point of view this device is quite recommendable for people that aren´t experienced in this hobby because it combines the advantages of the perfected drum tuner system and the modified output unit, that is much easier to handle. It took four whole years - that is from 1976 to 1979 - to produce the Satellit 2100, which was the sixth model of this series. A used device costs about $200 USD. Mint with case of course more.

Satellit 2100
Above: First edition. Below: the „Profi-Look“ edition.

Satellit 2100
Last edition looks like a Sat. 3000

Satellit 2100
Above: Rear view of „Profi-Look“ edition.
Below: Earphone „203A“

Satellit 2100
Dial cords, also fit for Satellit 2000

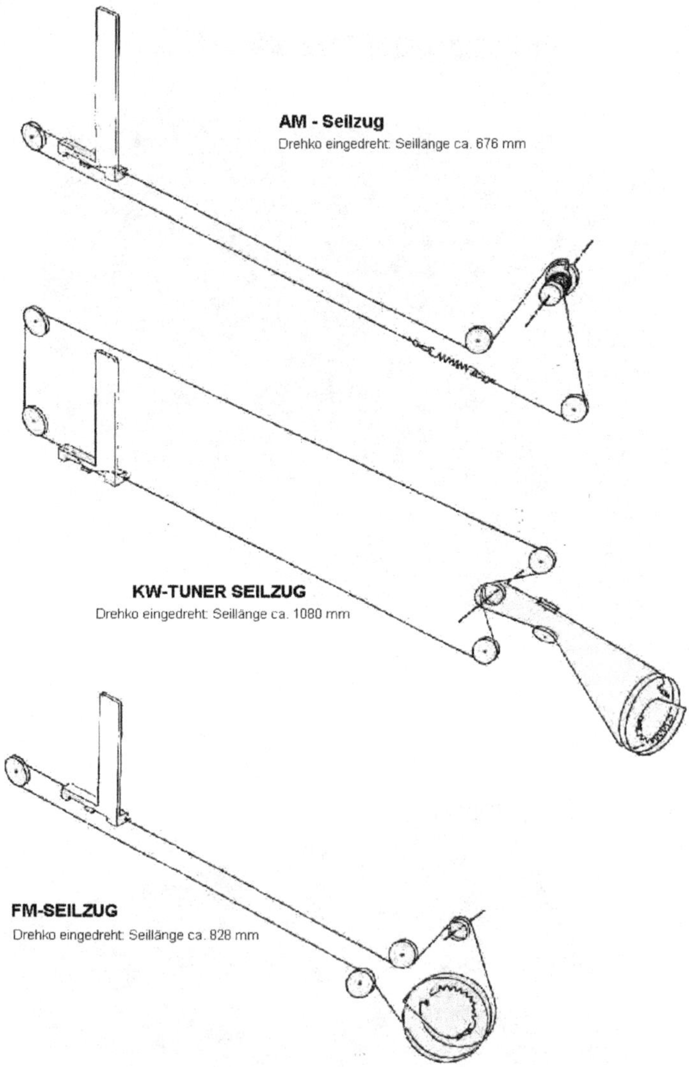

Satellit 3000

Build	1977 to end 1978
Dimensions	50 x 29 x 12 cm
Weight	8,9 kg
Body	plastic (black)
Ranges	FM, MW, LW, SW
SW-Bands	10, 8 of it at the Drum
SW-Range	160 to 10 mtr
Drum-Tuner	52 to 16,5 mtr
	double conversion
Bandspread	yes
Band With	three
Display	Digital, analogous and Drum
Output-Power	2,5 Watt Battery-operated
(Sine-power)	5 Watt Main-operated
Aerial (Telescope)	13 Elements
Powerpack	build in
SSB	build in
Clock	yes (inserted)
Timer	no
Memory	6 for FM
Dial illumination (Battery operated)	by troggle-switch
Connection-Sockets	1 DIN Socket (Line in/out) Earphone 9 V external Power Pack External Aerial
Accessories	Carrying Case Earphone 203 A
New price	DM 1.298,-

From 1977 on the model 3000 was offered for DM 1298, parallel to the S 2100. The most important innovation was the introduction of a digital frequency display. This display was working to one kHz and could be turned off to suppress interference caused by itself. Nevertheless, the three tuneable scales were maintained. Obviously, an abrupt change to a portable radio without any scales would have been too daring in those days. An SSB unit was placed instead of the FM scale, which was then put above for constructional reasons. The first intermediate frequency of the SW tuner got a quartz crystal to improve the strong signal resistance and the cross modulation. On the front a removable digital clock with a small LCD display supplied by button batteries was integrated between the signal strength readout and the digital frequency display. It was possible for DX fans and SW listeners to improve the reception by numerous tuning possibilities. On the back there were tuning controls to store six FM stations. This satellite type was the first to carry the SSB unit as standard. Still the drum tuner principle and the SW band spread was maintained, also the necessary band spread switch. Anyhow the first step from analogous to digital working was done. The amplifier unit, that had been taken over almost entirely from the S 2100, provided 2.5 watt during battery supply and 5 watt sine output during power supply. The Dryfit 476 accumulator could be used also instead of the six D-size batteries. Till that time the bodies of the satellites had constantly become bigger and bigger. That was also the case with the S 3000. A radio of that size can definetely be called a portable station receiver. Its outfit, however, seemed quite baroque. The loudspeaker panel, with its 40 small fields that looked like a waffle, was an example for that. This might have been one of the reasons why this radio was quite quickly replaced by the S 3400. Very last series has been made in a copper color. Those are much rare, even owners of a 3000 "cooper" might not know about that. Collectors are willing to pay about $400 USD for devices that are as good as new, but this price will probably even rise in the years to come due to the low number of pieces.

Satellit 3000
Digital frequencies counter built-in. A novelty shortwave listeners long waited for.

Satellit 3000
Photo above: Loudspeaker cover like a wafer
Photo below: Built in clock

Satellit 3000
Here is shown the operation section for SSB/CW.

Digital frequency counter: 7 segment unit (LED) with a resolution of ±1 kHz AM or ±10 kHz on FM band.

Satellit 3000
Frames, knobs and switches.

Satellit 3400

Build	1979 to end 1982
Dimensions	52 x 32 x 14 cm
Weight	8,9 kg
Body	plastic, black or brown
Ranges	FM, MW, LW, SW
SW-Bands	10, 8 of it at the Drum
SW-Range	160 to 10 mtr
Drum-Tuner	52 to 16,5 mtr
double conversion	
Bandspread	yes
Band Width	three
Display	Digital, Analogous and Drum
Output-Power (Sine-power)	2,5 Watt Battery-operated 5 Watt Main-operated
Aerial (Telescope)	13 Elements
Powerpack	build in
SSB	build in
Clock	yes (inserted)
Timer	no
Memory	6 for FM
Dial illumination (Battery operated)	by toggle-switch
Connection-Sockets	1 DIN Socket (Line in/out) Earphone 9 V external Power Pack External Aerial
Accessories	Carrying Case Earphone 203 A
New price	DM 1.349,-

This was an old device with a new design, what actually was the most essential difference to the Satellit 3000. The device seemed now tidier and more professional. About 200 devices a day were produced. The new design of the loudspeaker panel, for example, caught the eye immediately. This black lattice was much more elegant than the waffle of the S 3000. The three scales and the displays for frequency, time and tuning that had been integrated plainly were easier to clean than the dust-catching chinks of the previous model. Besides, the two handles on the left and right of the body gave it a somewhat "military look" and could serve as a protection if the radio fell forward. The entire body was black with a coarse meshed loudspeaker grill. The second series got a silver ring centered in the grill, only the last series were available in dark brown. The technical aspects were identical with the preceding model, i.e. a drum tuner with eight spread SW bands and the three already known scales that were illuminated when used. These were also known as floodlight scales. The four round toggle switches for the automatic frequency control (AFC) on FM, the battery testing mode, the band width switch and the SW band spread at the front of the device were replaced by more modern looking rectangular ones. All the switches and controls were then standardised and finished up the professional look. The figures of the then illuminated digital clock were enlarged and again a triple band spread switch on AM became possible. The "Super Wide" position was used for strong local stations to reach an optimum of fidelity. The S 3400 with its 8.9kg and the S 3000 were actually the heaviest devices of that series. It remains to be seen whether this fact fits in with a "portable" radio with battery box or not. Anyway, many DX fans generally used their devices at home. The price of a new Satellit 3400 was at about DM 1,350. It was in the shade of the Satellit 600 that had been promoted too early. Nowadays collectors already pay up to $450 USD for devices in good condition. In foreign countries some of used radios have even been sold at the price which had to pay in that period to get a new one. It might be advisable to follow internet auctions, although if much 3400's has been mixed and matched together by some German seller...

Satellit 3400
Black series

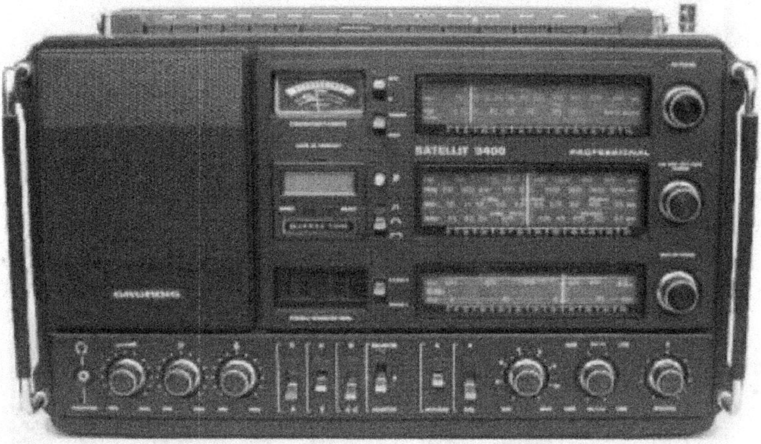

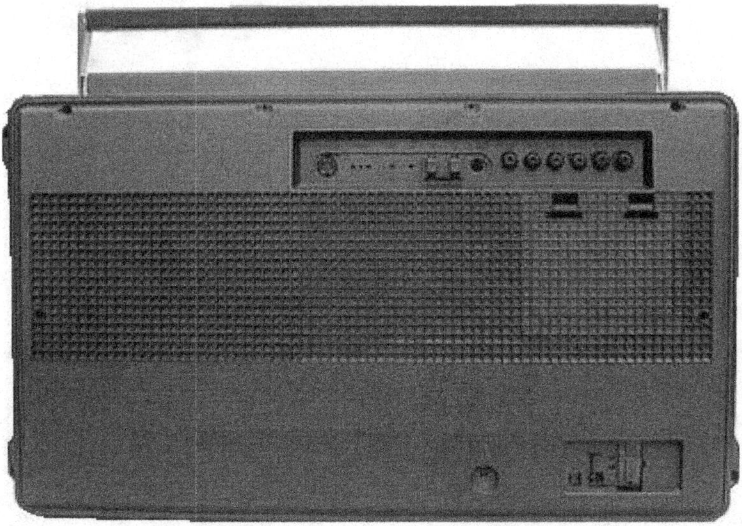

Satellit 3400
Separtely available: The original carrying case for Satellit.
Photo below: Coarse meshed loudspeaker grill of first series.

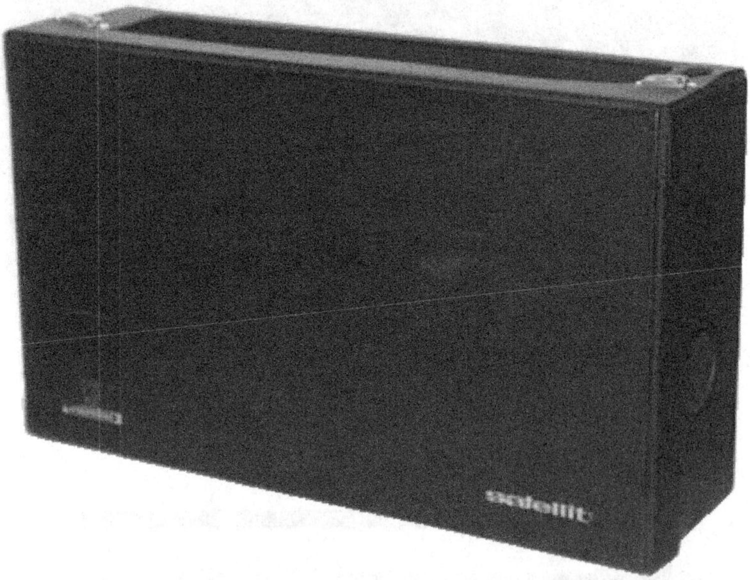

Satellit 3400

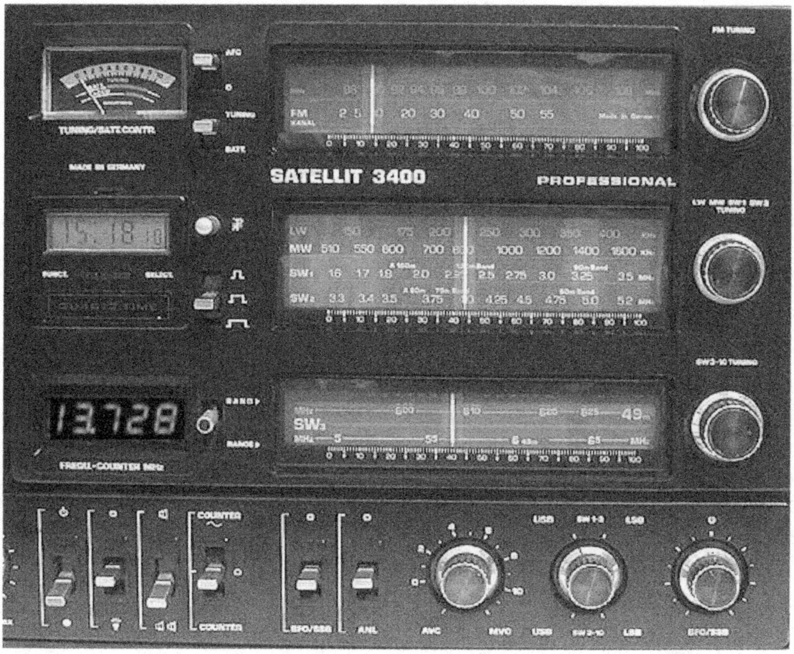

What's new? A bigger display for the clock, also middle position of the bandspread switch is marked.

Satellit 3400
Dial cords, also valid for Satellit 3000.

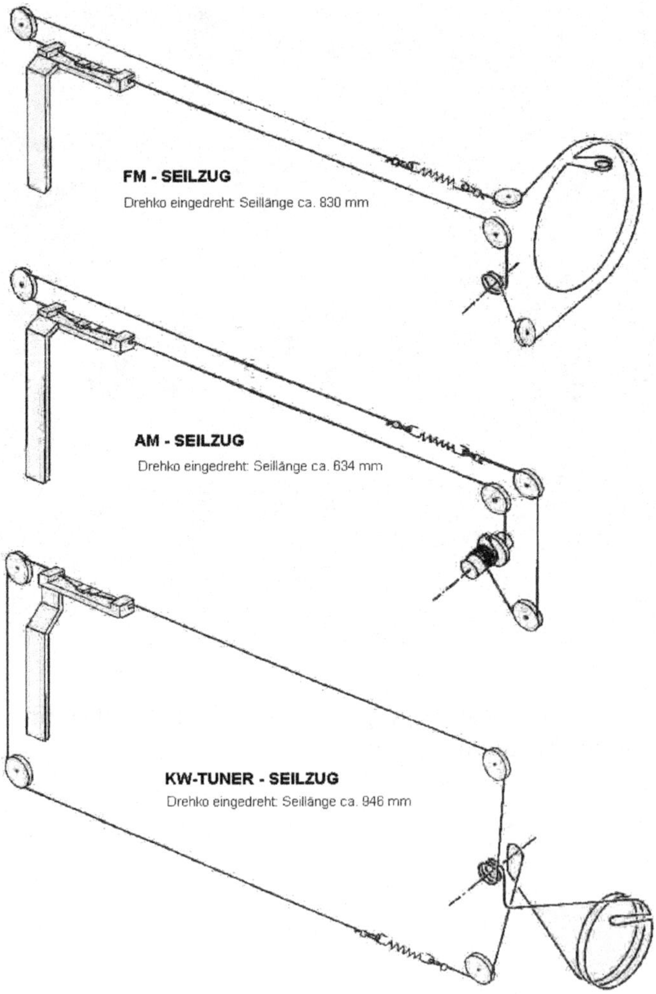

Satellit 1400

Build	1979 to end 1982
Dimensions	42 x 27 x 12 cm
Weight	5,5 kg
Body	plastic, black or brown
Ranges	FM, MW, LW, SW
SW-Bands	6
SW-Range	160 to 11 mtr double conversion
Drum-Tuner	no
Bandspread	no
Band Width	one
Display	Analogous and Digital
Output-Power (Sine-power)	2,5 Watt Battery-operated 4 Watt Main-operated
Aerial (Telescope)	13 Elements
Powerpack	build in
SSB	build in
Clock	no
Timer	no
Memory space	none
Dial illumination (Battery operated)	by toggle switch
Connection-Sockets	2 DIN Sockets (Line) Earphone 9 V external Power Pack External Aerial
Accessories	Headphone GDH series
New price	DM 648,-

In 1979 Grundig released three new radios called "Satellit". One of these was the Satellit 1400 that actually wasn´t supposed to be given the prestigious name of "Satellit". It was the smallest and cheapest model of that time. The cover of the instructions of use was completely yellow, the radio was black, with model 82 the colour changed to dark brown. A Grundig satellite without a drum tuner must have been hard to imagine for engrained fans of these radios, just as a Volkswagen with front motor would have been in the sixties. The relatively small radio was well equipped, namely with a digital and illuminated frequency display, with double conversion in the SW range, an SSB unit, a DX / local switch, an aerial trimmer with a scale reaching from 0 to 9 and a coaxial socket for the aerial. It was unpractical that the latter three elements had been installed on the back of the device. The German normed TA / TB sockets for gain supply, on the contrary, were quite beneficial, although also installed on the back. During power supply the output power unit provided 4 watt, during the use of six D-size batteries or an accumulator 2.5 watt. The power pack was already included, so that only a conventional cable for power supply was needed to turn the radio on. The cable could be stowed away for transport in a box on the back. The telescopic aerial had 13 elements and, as usual, there was a mark for FM reception at 81cm. If fully drawn out, the aerial was 142cm long and necessary for SW reception. The radio disposed of a so called double turn button for the tuning. The bigger outer ring was used for a fast tuning, the inner ring was needed for the fine tuning. As the following example for single side band function (SSB) shows, it was quite complicated to handle. First the upper range key had to be tuned to SW, then the lower one to the SW band one wanted to. After that, a toggle switch had to be laid to USB / LSB. Finally the frequency could be fine tuned by turning the inner part of the turn button. At that time a new radio cost DM 650, today collectors have to pay about $200 USD to get an S 1400. In the US the price is even higher.

Satellit 1400

Satellit 1400

Similar to the Oceanboy 820 (below). Both made in the early 80s.

Satellit 1400 (also satellit 2400)
Uncased AC transformator: all three fuses were accessible.

Satellit 1400
Left: Uncased transformator with plug
Right: The fixing holder with inserted anti-shock-gum

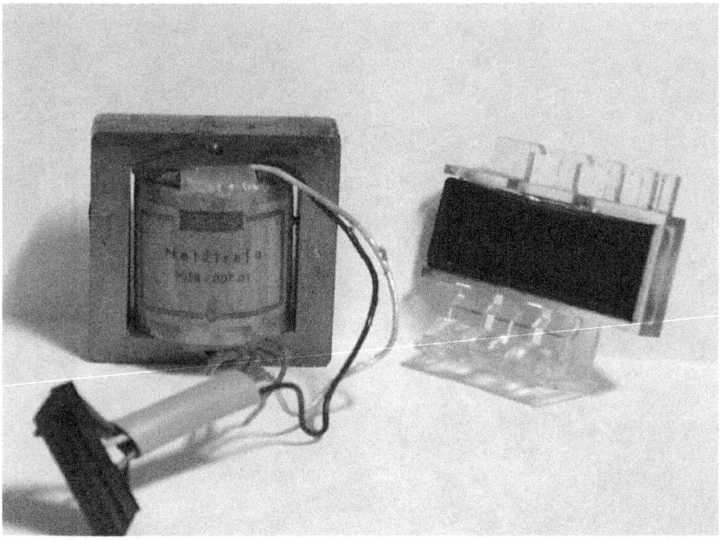

Photo below: Two DIN sockets and the open AC main cord case.
To clean the rear was a little bit difficult.

Satellit 2400 (Stereo)

Build	1979 to end 1982
Dimensions	53 x 29 x 12 cm
Weight	7,5 kg
Body	plastic, black or brown
Ranges	FM, MW, LW, SW
SW-Bands	6
SW-Range	160 to 11 mtr double conversion
Drum-Tuner	no
Bandspread	no
Band Width	one
Display	Analogous and Digital
Output-Power (Sine-power)	2x2,5 Watt Battery-operated 2x4 Watt Main-operated
Aerial (Telescope)	13 Elements
Powerpack	build in
SSB	build in
Clock	no
Timer	no
Memory space	6 for FM
Dial illumination (Battery operated)	by toggle switch
Connection-Sockets	2 DIN Sockets (Line) Earphone 9 V external Power Pack External Aerial
Accessories	Headphone GDH series
New price	DM 798,-

The second innovation of the year 1979 was the Satellit 2400 Stereo. As its name already reveals, it was a SW receiver with two integrated loudspeaker boxes and it could be used for listening to FM or to external signals in stereophony. It was up to each person to decide, whether to number such a device among the other SW receivers or not to do so. But it should be mentioned that till today such concepts, just as the ones with an integrated cassette recorder, haven't been paid much attention by DX fans and SW listeners. Yet a single glimpse reveals that a second box was added to the S 1400 model. Only if we take a closer look, we'll recognise some more differences in the appearance. The Satellit 1400 had a small grip in the form of a U, that could be hidden in the upper side of the device, so the radio also was quite nice to look at. Just as it had been the case with all the previous satellites, the Satellit 2400 was still equipped with the attached console. The grip also reminds strongly of the preceding models. In this console eight press keys were integrated, moreover six keys to select the FM station, a key to switch over to the FM scale and a key for FM function. The following options had to be given if you wanted to listen to a radio station on the 25 mtr. SW band: the AM key of the console on the upper side had to be pressed, the upper range switch had to be tuned to the short wave ranges S 2 – 6 and the lower range switch to S 3. In addition to that the toggle switch had to be set on USB or LSB to work with the single side band and - if necessary - the automatic gain control had to be switched to manual use. It´s no wonder that such switching lead to much criticism in the trade press after some tests had been realised. The voluminous appearance, due to the two loudspeaker boxes, also didn't find everyone's approval. Apart from that, technically it was identical with the Satellit 1400, also as far as the construction time is concerned. The body was available in black, from autumn 1981 on also in dark brown. The price of a new Satellit 2400 was DM 798. Nowadays collectors are willing to pay $350 USD for used devices in good condition because of the low number of pieces.

Satellit 2400

Satellit 2400
Dial cord, also usable for Satellit 1400
Photo below is: 1982 Grundig catalogue
„Radio and Recorder“ Rechargeable accu-pack

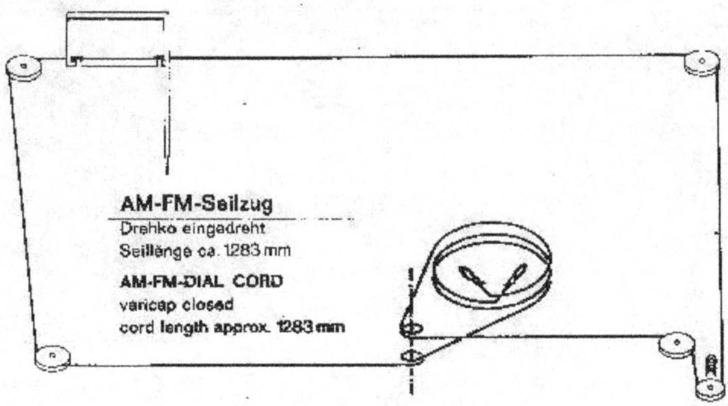

Satellit 2400
Rear, the loudspeaker are mount in different positions

Satellit 2400, above black vs. brown (below)

Satellit 4000 (Stereo / Rec.)

Build	1983 to end 1985
Dimensions	57 x 27 x 13 cm
Weight	7 kg
Body	Plastic, anthracite
Ranges	FM, MW, LW, SW
SW-Bands	2
SW-Range	75 to 13 mtr single conversion
Drum-Tuner	no
Bandspread	no
Band Width	one
Display	Digital
Output-Power (Sine-power)	2x2 Watt Battery-operated 2x4 Watt Main-operated
Aerial (elescope)	8 Elements
Powerpack	build in
SSB	no
Clock	yes
Timer	yes
Memory	35, 9 for FM, 18 for SW 4 for MW and 4 for LW
Dial illumination (Battery operated)	by slide switch
Connection-Sockets	1 DIN Socket (Line in/out) Earphone 110 / 220 V Main Power External Aerial
Accessories	Headphone GDH series Microphone (s) Cleaning Tape 461
New price	DM 849,-

What an adventure to integrate a cassette recorder! In 1983, again some devices of the Grundig Satellit series were replaced. The production of the S 1400, S 2400 and S 3400 was stopped. These models were replaced by three new ones, i.e. the small single conversion Satellit 300, the big S 600 and the Satellit 4000 Stereo - CR. Basically, this device could be considered a "grandchild" of the Concert Boy 4000 Stereo produced in the year 69. It was a good FM radio with two integrated loudspeaker boxes, each one of these boxes with a wide band loudspeaker and a tweeter. Besides that, it had an AM reception unit for MW, LW and SW up to 22,000 kHz. The sound was to be handled by slide controls again. At that time press keys and flat switches simply had to be part of a modern radio. Just as the Satellit 300 it was a device completely without any analogous scales. This was the first and only Satellit that combined both a radio and a cassette recorder in one. Till today this combination hasn´t been accepted because it represented more or less a compromise. The drive mechanism with its three motors could be controlled through electronical press keys. It was possible to record – also in stereo – radio and external signals or by using the two integrated microphones. The essential characteristics of the cassette recorder were its mechanical counter, an infinitely fade in or out while recording and an automatic rewind function after the play back. The possibility to switch over to "super stereo" was fine. The stereophonic sound was intensified through an electronical switching. The integrated 24–hour-timer made it possible to pre-program a recording, what actually depended on the running time of the tape. On the back of the device were two big openings with hinged lids. Into the right box six D-size batteries for battery supply could be inserted, into the left one three set-up cells, as well as the power supply cable. The output power reached 2 x 2 watt if batteries were used respectively 2 x 4 watt during power supply. The sound, however, was enormous. A new device cost DM 850. Although it was only a radio recorder with SW reception it is very demanded among collectors! Some of the devices might be available on common fleamarkets at a very low price, but there you should check the drive's function first.

Satellit 4000

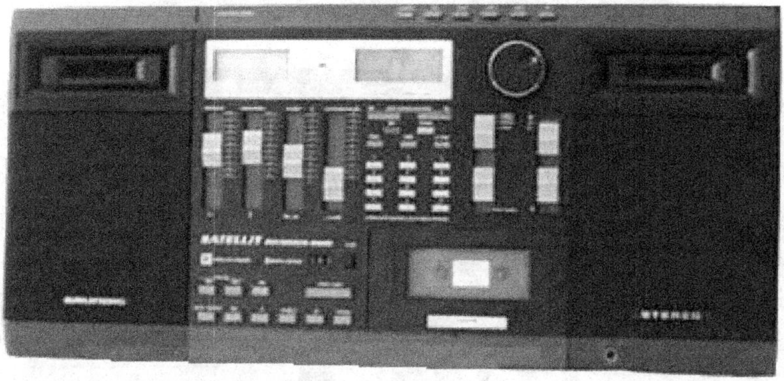

Satellit 4000
Below: The 1969 Concert Boy 210 (Transistor 4000).
Some features are common with Satellit 4000 stereo.

The table below is: A piece of Grundig 1984 catalog.
Satellit 400 a radio-recorder only?

Radio-Recorder

	Party-Center 2000 HiFi[1]	***Party-Center 1000***	***Satellit Rec. 4000***	***RR 3600***
Ausgangsleistung *(Watt)* ***bei Netzbetrieb*** *(Musikleistung)*	*2x25*	*2x20*	*2x7*	*2x10*
bei Batteriebetrieb *(Nenn- — Sinusleistung nach DIN 45324)*	*2x2,5*	*2x2,5*	*2x2*	*2x2*
Lautsprecher	*2 Boxen*	*2 Boxen*	*4*	*4*
Wellenbereiche	*U, K, M, L*	*U, K, M, L*	*U, 2xK, M, L*	*U, K, M, L*
Stationsspeicher	*10*	*6xU*	*9xU, 9xK1, 9xK2, 4xM, 4xL*	*6xU*
UKW-Scharfabstimmung (AFC)	–	●	–	●
Batteriekontrolle/Abstimmanzeige	●[2]	●[2]	●[2]	–/●

Satellit 4000

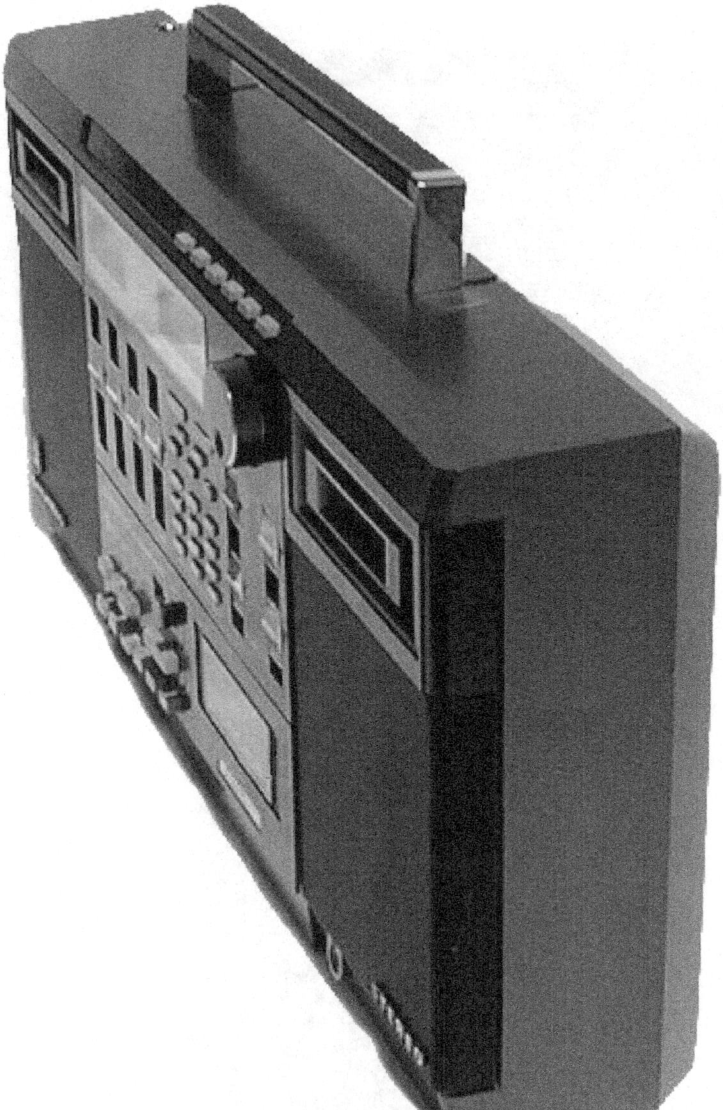

Satellit 4000

Photo above: Rear, central positionized - the DIN-socket

Photo below: Left battery compartment for setup batteries

Satellit	**600**
Build	1983 to end 1985
Dimensions	50 x 24 x 20 cm
Weight	8,5 kg
Body	plastic, black
Ranges	FM, MW, LW, SW
SW-Bands	no more divisions
SW-Range	160 to 11 mtr double conversion
Drum-Tuner	no, direct key input
Bandspread	no
Band Width	three
Display	Analogous and Digital
Output-Power (Sine-power)	2,5 Watt Battery-operated 8 Watt Main-operated
Aerial (Telescope)	13 Elements
Powerpack	build in
SSB	build in
Clock	yes
Timer	yes
Memory	60, 16 for FM, 34 for SW 4 for LW and 8 for MW
Dial illumination (Battery operated)	by toggle switch
Connection-Sockets	1 DIN Socket (Line) Earphone Line-in and Line-out External Aerial Tape control (auto-start)
Accessories	Headphone GDH series
New price	DM 1.398,-

1983: After some experiments Grundig brought another great radio on the market. The Satellit 600, that replaced the Satellit 3400, was a completely new device. First time it was possible to tune the pre-selection separately. This could be manually or by the automatic. Although the size of the body didn´t really impress much through elegance, the chassis could still be used even if the body was completely removed. This was quite beneficial for the production process and if repairs had to be done. Many of the technical details had actually been well thought out. Gums were used for the bearing of the power transformer, whereby it was kept in an almost floating position. Six D-size batteries or the well known accumulator 476 were necessary if you wanted to use the Satellit as a portable radio. For the first time it was then made possible to enter the frequency directly through a keyboard with ten keys. This not quite cheap concept was therefore introduced in all the three new releases, i.e. not only in the S 600 but also in the S 300 and the S 4000. Thereby, it could finally be dispensed with the need of a drum tuner, but not with the need of an analogous scale. The keyboard could be opened from the outside, what actually proved to be favorable under certain circumstances, e.g. if the illumination had to be exchanged. The device had a number of useful extras, such as a triple switching for the AM band width, the possibility to control the input sensitivity on AM and 60 memories, 32 of these were thought for SW use. After the AM range frequency had been entered, the preselector and the dial indicator followed, driven by an electronically complicated motor system. This was realised through toothed wheels, pinions and tooth racks in the newly developed variometer assembly. Nevertheless, this unit could be turned off to make the pieces follow by manual operation. The different assemblies were connected with modern contact plugs. On the right side of the front an aerial trimmer for the tuning of an external aerial was integrated. A new device cost DM 1,400. Collectors are willing to pay more than $350 USD for a Satellit 600 in good condition, for a mint 600 even $500 USD. If you´d like to buy one, you should check if the motor control is working. 4500 units with a golden scale has been made in fall 1983. Approx. 25.000 units had a black scale (1984 / 1985)

Satellit 600

Satellit 600
Photo above: The variometer module
Photo below: Metal made switches

Satellit 600
Two lamps to illuminate the clock (see arrows)
Photo below: Two snaps are fixining the clock-unit

Satellit 600
Photo above: Timer and memory block
Photo below: Owner's Manual (1983)

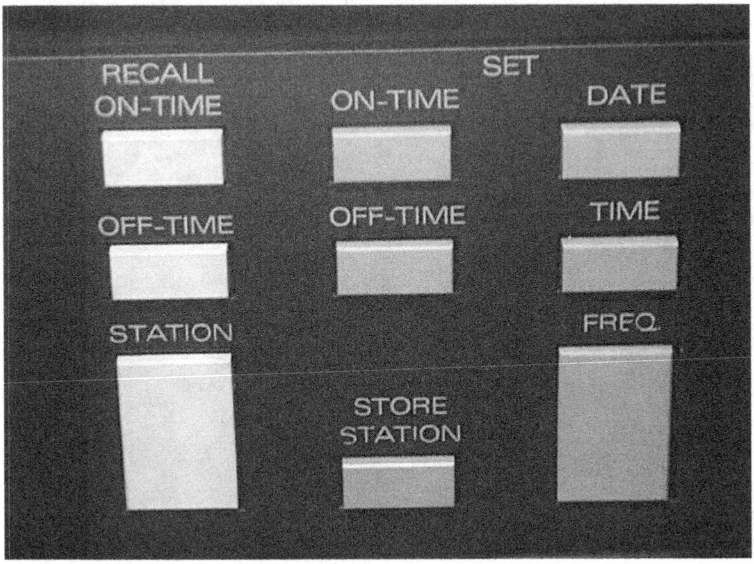

Satellit 650

Build	1986 to Dec. 1991
Dimensions	50 x 24 x 20 cm
Weight	8,5 kg
Body	plastic, black and grey
Ranges	FM, MW, LW, SW
SW-Bands	no more divisions
SW-Range	160 to 11 mtr, later to 10mtr
	double conversion
Drum-Tuner	no, direct key input
Bandspread	no
Band Width	three
Display	Analogous and Digital
Output-Power	2,5 Watt Battery-operated
(Sine-power)	8 Watt Main-operated
Aerial (Telescope)	13 Elements
Powerpack	build in
SSB	build in
Clock	yes
Timer	yes
Memory	60, 16 for FM, 34 for SW
	4 for LW and 8 for MW
Dial illumination (Battery operated)	by toggle switch
Connection-Sockets	1 DIN Socket (Tape)
	Earphone
	Line-in and Line-out
	External Aerial
	Tape control (auto-start)
Accessories	Headphone GDH series
New price	DM 1.498,-

The first devices of this series were „Grey-blue". International 650's code was:G-RV 57; older 650 Professional one's was: G-RV 56. Newer 650's (anthracite only) at fall 1989 got the code: G-RV 58-51 P. Besides that, they didn´t show any big differences to the S 600 in their design, except that the possibility to connect a DF aerial for navigation in the LW range was given. The socket was installed on the front of the device just as the switch to control the aerial. It stays open why the tuning control of the external aerial for the radio bands wasn´t taken over from the S 600. After the licensing regulations in Germany had been loosened in autumn 1988 the radio was also available with an extended SW unit for up to 30,000 kHz on the German market since fall 1989. In 1991 still the name of the Satellit 650 appeared in receiver tests and buying advises of trade magazines, even though the radios were then lovingly called "modern baroque". This receiver was offered simultaneously to the S 400 and its successor, the S 500. The fact that none of the two modern radios S 400 / S 500 became that popular the Satellit 650 had been was remarkable. In October 1991 Grundig decided to take the S 650 out of the programme. With the then presented Satellit 700 a more efficient but considerably smaller and lighter receiver was put on the market. Moreover, the price of the Satellit 650, that was DM 1,498, was much higher than the price of the Satellit 700. In memory of the big satellites and as a last chance to buy one, Grundig offered 1,000 devices as a final edition from autumn 1991 on. With each of these 1,000 devices a certificate indicating the current edition number, technical and historical documents about the Satellit series as well as a pair of headphones were enclosed. The current and certificated edition number was also engraved on the body. The "code" engraved in the bottom of the body was here G-RV 58 51 AB-P. This one was the most expensive Satellit of all the series with its new price of almost DM 1,500. Today the 650 receiver is still in use in many places. That's why nowadays the value of a used one respectively the price of a collector's item is at about $400 to $550 USD. Yet some of the devices "Final Edition", that were in mint condition, have changed hands for US $ 800.

Satellit 650

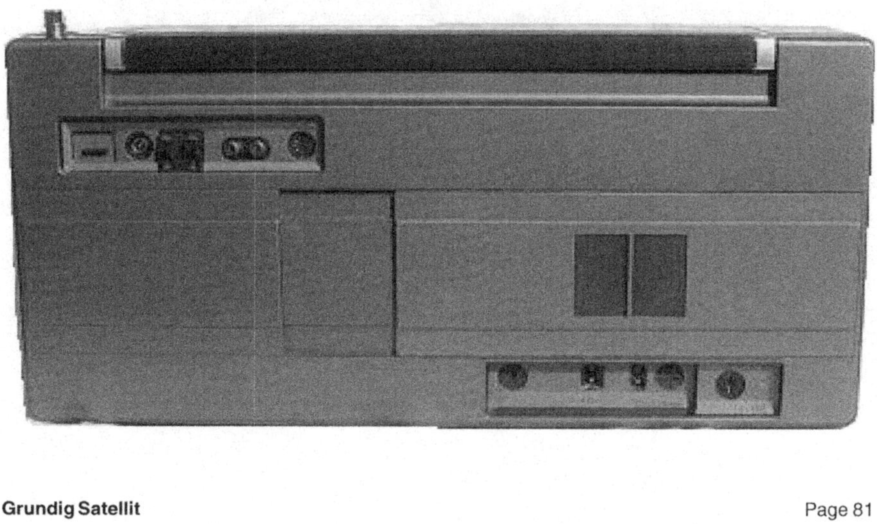

Satellit 650
Upper pic: Until fall 1988 radio grey-blue in color.
Picture below: Another design plus the antracite color indicates a new series (1989 to 1991).

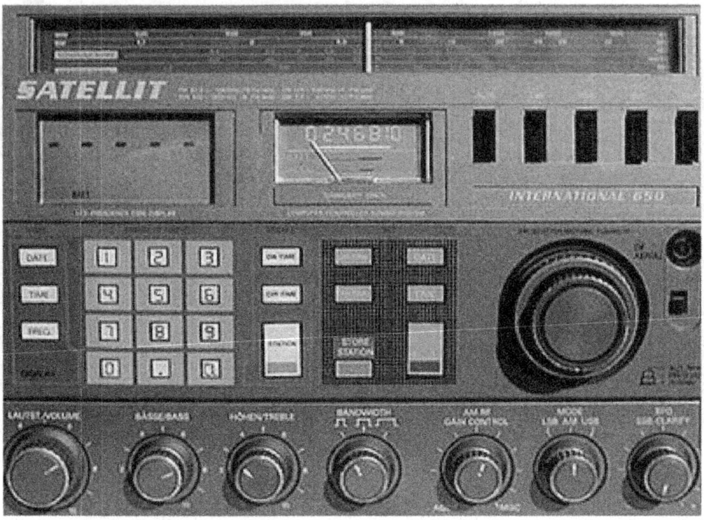

Satellit 650
Removed body
Photo below: satellit 650 Final Edition, here #400

Satellit 650
Connection sockets at the rear

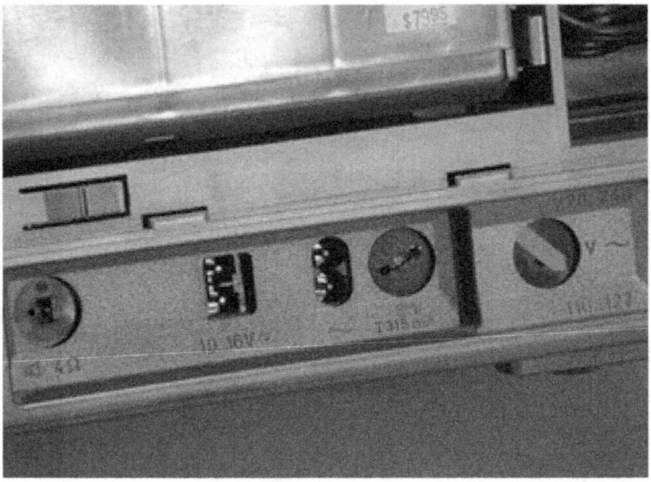

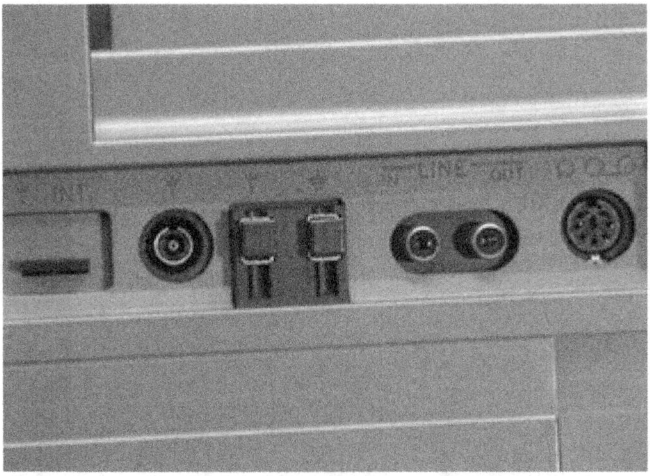

Satellit 300

Build	1983 to end 1985
Dimensions	31 x 18 x 7 cm
Weight	2,1 kg
Body	plastic, black or brown
Ranges	FM, MW, LW, SW
SW-Bands	2
SW-Range	75 to 13 mtr single conversion
Drum-Tuner	no, direct key input
Bandspread	no
Band Width	one
Display	Digital
Output-Power (Sine-power)	2 Watt Battery-operated 2 Watt Main-operated
Aerial (Telescope)	8 Elements
Powerpack	build in
SSB	no
Clock	yes
Timer	yes
Memory	35, 9 for FM, 18 for SW 4 for LW and 4 for MW
Dial illumination (Battery operated)	by toggle switch
Connection-Sockets	1 DIN Socket (Line) Earphone External Aerial 110 / 220 V Main Power
Accessories	Headphone GDH series
New price	DM 450,-

This digitally controlled device reminds strongly of the Yacht Boy 700. The very first units has been made in silver with cromiumplated buttons. Probably Grundig marketing dept. understand both radios are looking too similar? That might be the reason, the Satellit 300 got's quick another color - black. But the S 300 had no band width switch and no SSB unit to receive single side band transmissions. Nevertheless, compared with the battle-ships 3400 respectively 600 it was a cheap alternative. The radio could store 35 frequencies and had a programmable timer function to turn the set on automatically. Sound and volume had to be regulated by slide controls again! The integrated power pack or six C-size batteries had to be used to supply it. To store the data, three supplementary AA-size batteries were needed. The device could be tuned by using a turn knob, by entering the frequency directly through the keyboard or through a scan mode. The direct frequency entering however was something you had to get used to. First you had to chose the wave range through a press key, then to press the "Set" key. After that, the frequency had to be entered through the keyboard. But still that was not enough! Finally the "Freq." key had to be used to activate the set. The wave ranges and the display illumination were operated through seven tap keys on the upper side of the device. The eight-piece aerial was no Satellit aerial, but proceeded from the Yacht Boy. Though the 2 watt sine output power seemed little, it was sufficient to listen to all radio bands. A new item cost DM 450 and was considerably cheaper than the S 3400 / S 600 had been. But it's certainly true that this radio was not more like a Yacht Boy, even the modified 300a with better AM-IF Filter. A Grundig quotation: "We have learned from the mistake we made with the S 300." Today it's difficult to find a Satellit 300 on the second hand market. For a used one in good condition $250 USD or more should be taken into account. An mint 300 could cost you $400 USD. As an advice, some of the devices might be sold on fleamarkets for $30 because their appearance reminds rather of cheap supermarket radios.

Satellit 300

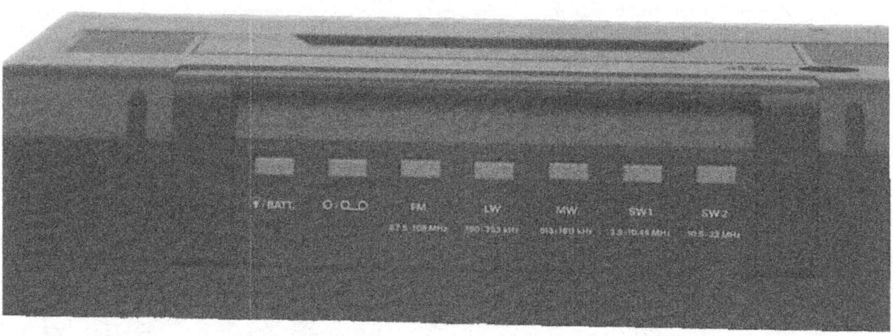

Satellit 300
Ultra-rare model in silver.

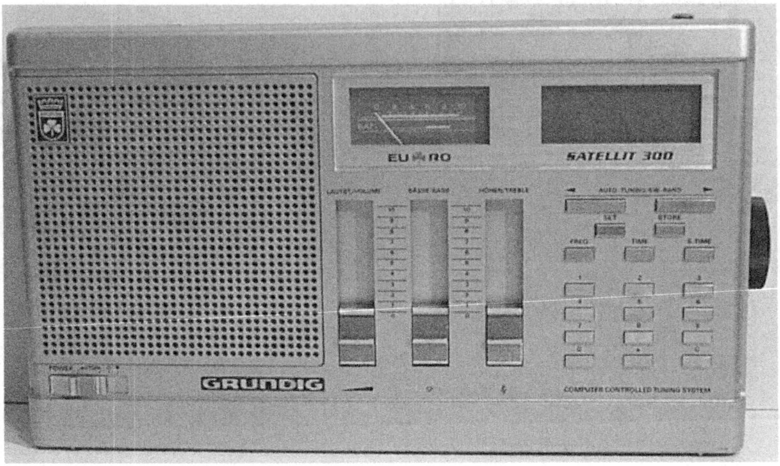

Photo below: Grundig Yacht Boy 700

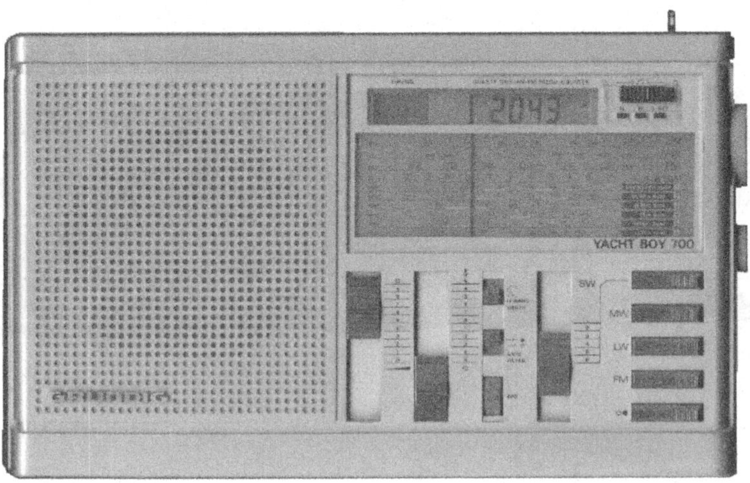

Satellit 300
Convenient operation by tilted position

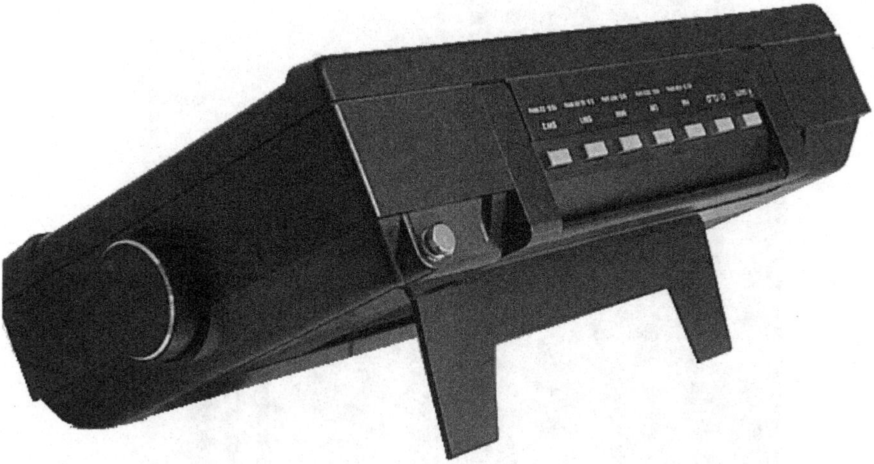

Satellit 300

Satellit 400

Build	1986 to end 1988
Dimensions	31 x 18 x 7 cm
Weight	2,15 kg
Body	plastic, grey
Ranges	FM, MW, LW, SW
SW-Bands	no divisions
SW-Range	160 to 11 mtr (later to 10mtr) double conversion
Drum-Tuner	no, direct key input
Bandspread	no
Band Width	one
Display	Digital
Output-Power (Sine-power)	2 Watt Battery-operated 3 Watt Main-operated
Aerial (Telescope)	11 Elements
Powerpack	build in
SSB	build in
Clock	yes
Timer	yes
Memory	24, free selectable
Dial illumination by key (Battery operated)	
Connection-Sockets	1 DIN Socket (Line) Earphone External Aerial External 9 V Power Pack 110 / 220 V Main Power
Accessories	Headphone GDH series
New price	DM 650,-

Compared with the previous model its appearance didn't show many differences, the body was almost identical. First of all it's the Grey-blue body that catches the eye, just as the subdued anthracite loudspeaker panel does. The Satellit 400 was (again) equipped with the "right" aerial that consisted in 11 pieces. However, experts know that this is only the small 118cm-long version. One technical improvement was e.g. that from then on the SW range started at 1,612 kHz and ended at 26,100 kHz. International series up to 30.000 kHz. To avoid image frequencies the double conversion principle was applied. Grundig had transferred the plug for an external aerial from the back to the right side of the device. To guarantee a comfortable use on the table, the fold away handle on the back could be pulled out, just as it had been the case with the preceding type. The S 400 disposes of an SSB unit. The tap keys for the wave ranges were placed on the front, what made the upper side of the device plain and free of control elements. The strong signal resistance left a bit to be desired and therefore the DX / local switch was added. One of the radio's characteristics is its clear front. Working with the device has become easier. Enter the frequency, then press the "Freq." key – done! The power pack was installed firmly, just as in the S 300. Yet, the device could be supplied "from the outside" by using a 9v power pack. Battery or accumulator supply was also possible. Six C-size batteries were then needed to supply the set and three AA-size batteries for the clock and the memories. The amplifier unit alone could be used through the German standard socket for line in / out combined with the AUX key. The 3 watt sine output power during power supply was quite respectable. Besides this, the user had the possibility to store a second time. All this made the set a usable radio that unfortunately had one disadvantage. It came years too late, because the numerous radios that were produced in the Far East caused a stiff competition. The price of a new one was DM 650, today's price of a used one is at about $250 USD, for one in mint condition with papers you even have to pay $400 USD. Beware if 400 Professional is running up to 30.000 kHz! Somebody opened and "modified" it

Satellit 400
Factual elegance made in Bavaria

Satellit 400
Owner's manual

Photo below: 9 Batteries in the Satellit 400

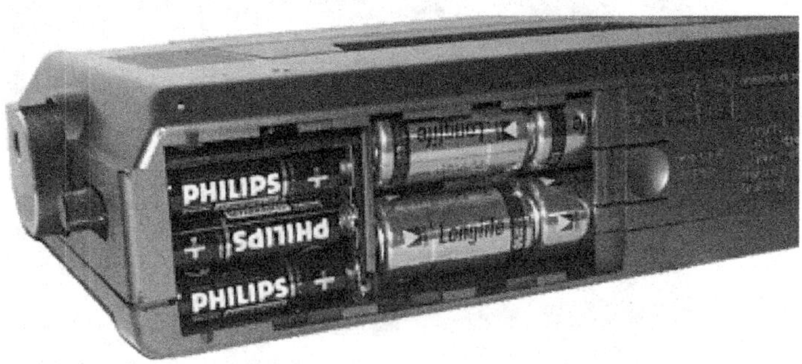

Satellit 400
Connection Sockets

Satellit 400
Rear

Satellit 500

Build	1989 to end 1991
Dimensions	30 x 18 x 7 cm
Weight	1,8 kg
Body	plastic, black
Ranges	FM, MW, LW, SW
SW-Bands	no divisions
SW-Range	160 to 11 mtr, later to 10 mtr
	double conversion
Drum-Tuner	no, direct key input
Bandspread	no
Band Width	two
Display	Digital
Output-Power	1 Watt Battery-operated
(Sine-power)	1,5 Watt Main-operated
Aerial (Telescope)	11 Elements
Powerpack	build in
SSB	build in
Clock	yes
Timer	yes
Memory	42, free selectable
	plus 156 at the ROM-Table
Dial illumination	automaticly by key input
(Battery operated)	keys also illuminated
Connection-Sockets	Line-out (Line)
	Earphone
	External Aerial
	External 9 V Power Pack
	External Loudspeaker
	Tape control (auto-start)
Accessories	Headphone GDH series
New price	DM 748,-

This radio was part of the Grundig program for two years, i.e. from 1989 to1991. It was the successor of the Satellit 400 and at the same time a cheap alternative to the S 650. Probably no DX fan had replaced the S 650 by a S 500! For the first time a multifunctional display was integrated. Thus it was possible to label the 42 memos alpha numerically on the display. The double conversion system disposed of a synchronous demodulator to suppress interference. Again Grundig had replaced the slide controls of the output unit by turn controls. Something quite similar happened with the power pack, because the producer didn't have an over all concept for this unit. The S 500 / S 700 were again supplied by the external NR 90 Grundig power pack. At the beginning, this series showed strong differences in quality. If you had a defective device, strong local FM stations could reach through the high frequency bands of the SW range. For barely DM 800 the SW listener got a modern radio with SW and SSB. The AM unit wasn't completely free of noise. A fine feature was the display illumination during battery or accumulator supply. After a new frequency had been entered, the illumination was turned on for a few instants. The smallest SW step was 100 kHz. The 1,5 watt output power was quite respectable. By using four D-size batteries it was possible to listen to the radio for up to 90 hours under way. A testing mode, see also chapter S 700, could be activated. The radio had a band width switch for AM use as well as a DX / local switch. The German normed output socket was replaced by a modern cinch plug, but unfortunately it could only be used for mono line out. Three 3.5mm jack sockets were thought for the use of headphones, an external loudspeaker and a cassette recorder control. The "Lock" control blocked the keyboard, if it was touched inadvertently. Since autumn 1990 the SW range reached up to 30,000 kHz, before it had ended at 26,100 kHz. A used device in good condition shouldn't cost more than $150 USD. For radios in a excellent shape, however, you might have to spend about $200 USD. Mint in box up to $350, more you shouldn't spend. Beware 500's with a label "Professional 26.100 kHz" at the bottom, while the radio is running up tp 30.000 kHz. This radio was opened and "extended" by somebody !

Satellit 500

Satellit 500
Owner's manual 1990, Photo below: Short-wave guide 1990

Satellit 500
Right: Satellit 500 see the small switches
Left: Satellit 700, bigger switches and clarify for SSB

Satellit 500
The original package

Photo below:
G-RY 56 stand for Satellit 500 Professional (German national market)
A radio with such a label should *not* tune up 30,000 kHz. If it does, it has been opened!

Satellit 700

Build	1992 to March 1996
Dimensions	30 x 18 x 7 cm
Weight	1,8 kg
Body	plastic, anthracite
Ranges	FM-RDS, MW, LW, SW
SW-Bands	no divisions
SW-Range	160 to 10 mtr, double conversion
Drum-Tuner	no, direct key input
Bandspread	no
Band Width	two
Display	Digital
Output-Power (Sine-power)	1 Watt Battery-operated 1,5 Watt Main-operated
Aerial (Telescope)	11 Elements
Powerpack	build in
SSB	build in
Clock	yes
Timer	yes
Memory	512 extendable to 2.048, plus 120 at the ROM-Table
Dial illumination	by slide switch (Battery operated)
Connection-Sockets	Stereo Line-out Headphone External Aerial External 9 V Power Pack External Loudspeaker Tape control (auto-start)
Accessories	Headphone GDH series
New price	DM 900,-

In October 1991 Grundig presented its newest product on the International Radio and Television Fair (IFA) in Berlin. Only at a second look one could distinguish it from the S 500. The S 700 had the same body and therefore the same size and weight as the S 500. But technically it was completely different. It actually replaced two devices, namely the Satellit 500 and the very successful Satellit 650. As standard it had 512 memories that could be extended up to 2,048 by inserting EEPROM memofiles. This one was the first portable radio with an RDS system for the FM range. On the AM range a preselector was working constantly (also Satellit 500), the telescopic aerial tuned to this preselector made an external aerial (almost) unnecessary. The synchronous demodulator was replaced by a synchronous detector, so if a side band showed interference, it could be turned off almost entirely. The extended FM range from 78,3 to 109 MHz, the memory functions and some menu controlled extra functions are only some of the characteristics of this device that was available since January 1992. The two switches for the aerial and the band width were enlarged and, during battery supply, the display illumination could be activated through a slide switch for a few seconds. When the device was turned off, only the left small glass socket light was burning faintly. To exchange the small light some sensitivity and patience is needed. The whole device can only be taken to pieces, if you start at the back. The lights are all soldered in. It seems as if one of the IC programmers wanted to perpetuate his memory by introducing his date of birth. So the testing mode can be activated by choosing the code "050251" and then pressing the "store" key, the device must be turned on to do so. If you chose the figures "123456" before turning the set on, the device is then to the exhibition mode and can't be used. By repeating this procedure the set can be tuned back to its normal function. In the MW range this radio from Bavaria shows all its abilities, especially the low noise in the power output unit might be mentioned in this context. The last series got little stronger LCD read-out, no dust cover and anthracite owners manual. Today used devices in good condition can be acquired for about $250 USD, mint in box all papers, then $400 and more.

Satellit 700

Satellit 700
Behind the removable hatch are slots for the three memofiles. To remove the memofiles, a small plastic-made pliers was added.

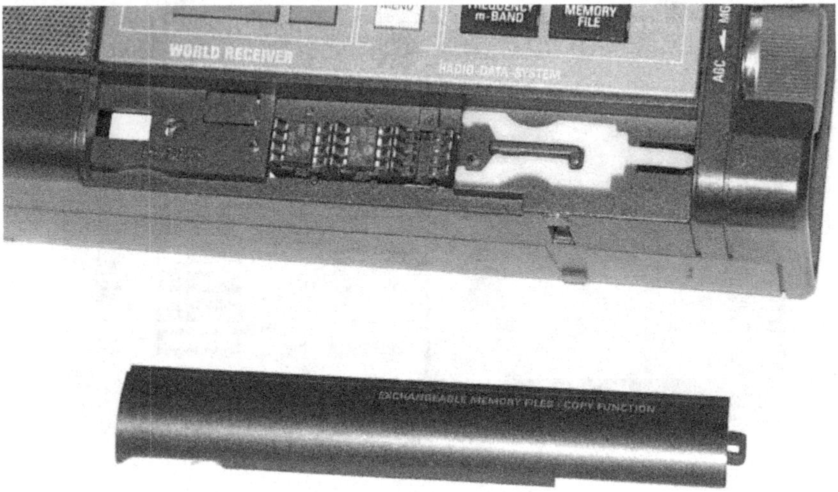

Photo below: The power pack NR 60

Satellit 700
Radio in „working“ position.

Satellit 700
To protect the radio and keep it clean from dust, a transparent cover was added in the first series, until 1994 only.

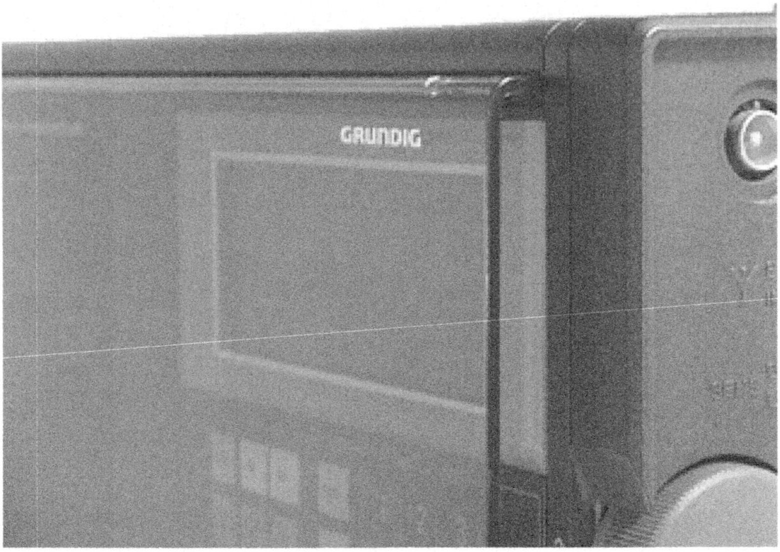

Satellit	**900**
Build	Summer 1995
Dimensions	33 x 18 x 6 cm
Weight	1,8 kg
Body	plastic, grey
Ranges	FM-RDS, FM-OIRT-Band, MW, SW, LW starts at 100kHz
SW-Bands	no divisions
SW-Range	160 to 10 mtr, double conversion
Drum-Tuner	no, direct key input
Bandspread	no
Band Width	three
Display	Digital
Output-Power (Sine-power)	0,5 / 1 Watt Battery-operated 1 / 3 Watt Main-operated second data in Boost-mode
Aerial (Telescope)	9 Elements, 100 cm
Powerpack	build in
SSB	build in synchron detector
Clock	yes
Timer	yes
Memory space	512 extendable to 2.048, free selectable plus ROM-Table
Dial illumination	by button (Battery operated)
Connection-Sockets	Stereo Line-out Line-in (mono) Earphone External Aerial External 9 V Power Pack External Loudspeaker Tape control (auto-start) Computer serial-bus
New price	DM 1.000,- (planned)

The planning of the Satellit 900 goes back to the year 1993. It took almost two years to develop it until finally, in August 1995, the device was presented by request on the IFA in Berlin, but it was presented with greatest caution, because at that time the CPU that had been delivered didn´t comply with the quality demands Grundig made. Therefore only ten finished devices have been produced of which nine have stayed at Grundig´s until today. It was planned to start selling the type in January 1996, but the date had to be postponed and was finally given up at the end of 1996. The CPU producer wasn´t able to meet the heavy demands Grundig made. The development and production of this small pre series had swallowed enormous sums, but a radio that only guaranteed a bad reception would have caused the Grundig group a loss of prestige and thus a much bigger damage. There would definitely be no S 900, except under licence if all the strict quality demands and criteria would be fulfilled. Some of the control elements proceed from the Yacht Boy 500, the store including the memofiles are compatible with the S 700. The display was replaced by a full DOT field with 160 x 24 pixels and was thought for frequency, station name, store, times, radio text etc. It had a size of 11.3 x 8.3cm. Compared with the S 700 the most important innovations were the triple band width switch on AM, the extended FM range from 65 to 108MHz, the Boost turn switches to increase the output power 3 or 4 times, a scrolling display for all the stores of a memofiles, the AM tuning in steps of 50Hz, a line-in plug as well as the possibility to use it as a computer interface. For the first time the aerial of a Satellit was on the left side of the body and it´s easy to explain why. The modern digital technology causes interference, even if combined with a big display. The farest point from this source was on the very left. The Satellit 900 was to cost DM 1,000 and a production of six years had been conceived. This radio was meant to built the bridge between new radio technologies, as e.g. DAB, and the old ones.

Publisher's note:
This model would eventually be brought to market in the United States as the Eton E1. A model E1 XM was also produced that could additionally receive the XM satellite subscription service with optional XM antenna module.

Satellit 900

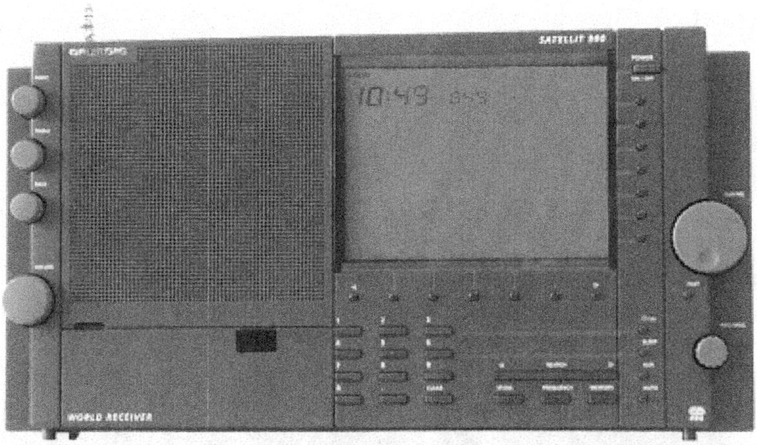

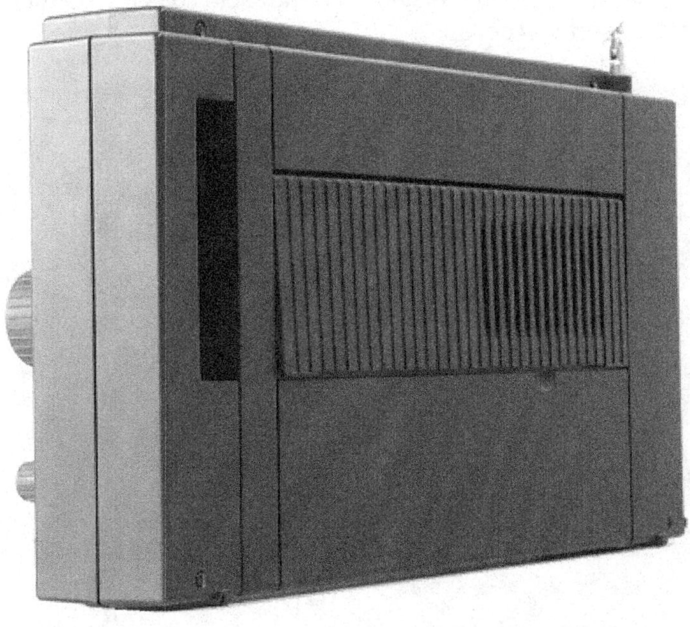

Satellit 900

Behind the (open) lid: Sockets for three memofiles, display contrast regulator, the reset-button and PC-serial connection socket.

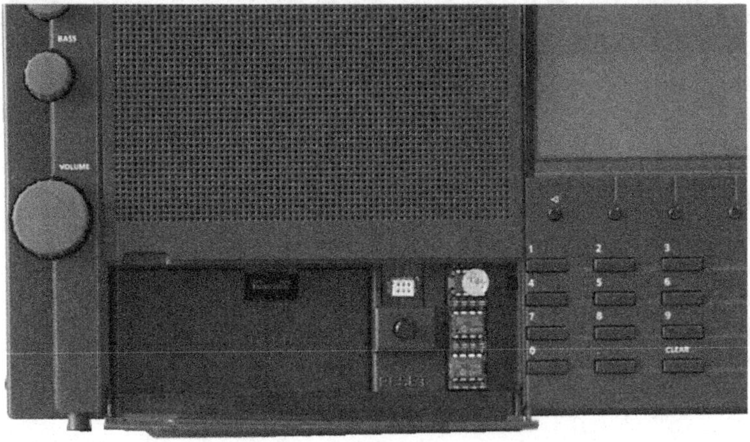

Photo below: Behind the second removable lid is the battery-case to insert 4 D cells.

Satellit 900
All connection sockets

Satellit 900
Huge multi-display

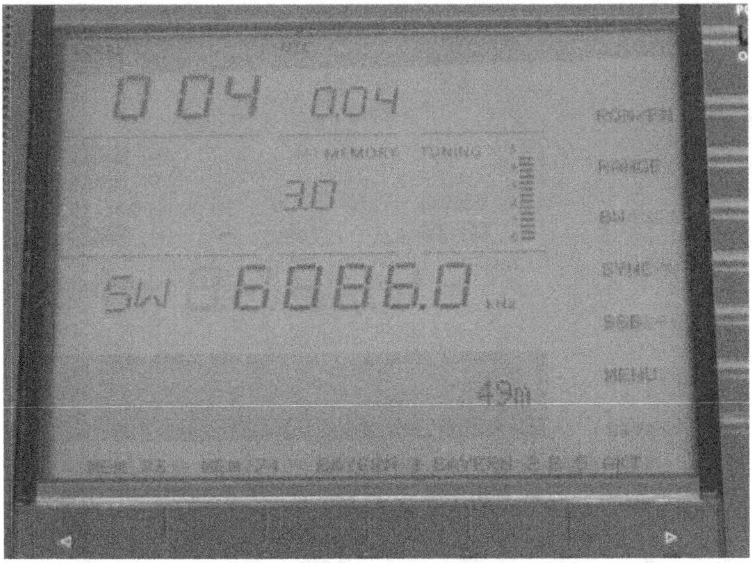

For external control, a software program was developed. It should be possible to operate the Satellit 900 by a computer.

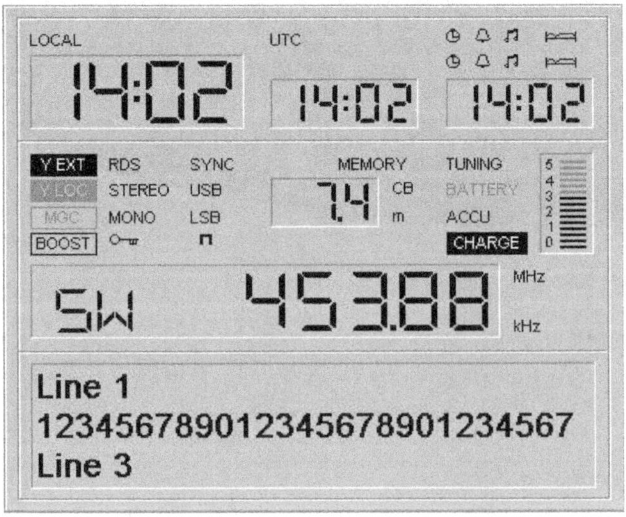

Satellit 800

Satellit	800
Build	End of 1999 to ?
Dimensions	53 x 24 x 22 cm
Weight	6,6 kg
Body	plastic, grey
Ranges	FM, VHF, MW, LW, SW
SW-Bands	no divisions
SW-Range	160 to 10 mtr double conversion
Drum-Tuner	no, direct key input
Bandspread	no
Band Width	three
Display	Digital
Output-Power (Sine-power)	2 Watt Battery-operated 2 Watt Main-operated
Aerial (Telescope)	13 Elements
Powerpack	external
SSB	build in
Clock	yes
Timer	yes
Memory	70, free selectable
Dial illumination (Battery operated)	by pushing lamp switch control
Connection-Sockets	Line Earphone 9 V external Power Pack External Aerial Aerial PL- Socket
Accessories	
New price	$499 USD (DM 1.050,-)

In fall 1999, some information about a new radio called Satellit 800 appeared in the trade magazines. Somewhat later this news started to take shape and could then also be found in the internet. In the year 2000 edition of the WRTH, the World's Radio and TV Handbook, the radio was offered officially for the first time at the price of US $ 499 on the US market. In April 2000 the radio, that is made in China and distributed by Lextronix– an US company, is to be sold on the German market. Nowadays, there are no current negotiations for a licence release. Its appearance fits completely in with the great Satellit series. The last model had been the S 650. The Satellit 800 disposes of FM, MW, LW, the SW range till 30,000kHz and the VHF range from 118 to 137MHz. The big LCD display shows all the functions and switch positions the device has. The S 800 can be supplied through an external power pack or six D-size batteries. There are 70 memos disposable and the analogous signal strength readout gives information about the strength of the radio waves that fall in or about the battery voltage. The loudspeaker and the ferrite antenna (AM) are a little bit tiny. The 142cm–long telescopic antenna is placed on the right side and seems to be same as of Satellit 600 / 650.The two bars on the front are meant to face down the device to open the radio. There are sockets for loudspeakers, FM and AM aerials, headphones and line-out. 3 band widths on AM can be chosen and thus guarantee an optimal listening. What is really good is that the filters have been chosen on 6kHz, 4kHz and 2.3kHz. The narrow 2.3kHz filter is especially beneficial while using the single side band. Unfortunately there is no option for manual gain control, which is needed for SSB. According to tests in US the 800 guarantee an excellent reception. The 800 has been tested by Grundig laboratory in Germany with unsatisfying results. Therefore no license for Europe! The operating guides are written in English, French and Spanish only, what indicates that this radio is thought for the American continent. While recently radios have rather become smaller, the size of the Satellit 800 is out of all proportion to any battery supplied portable radio. It's distribution in Germany is rather unlikely. Is it an Grundig?

Satellit 800
Will this Satellit, made in China, ever come to the European market?

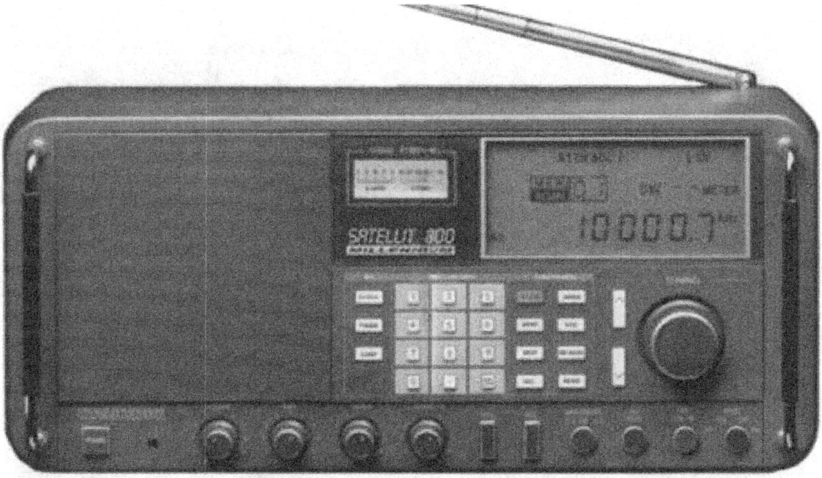

Satellit 800
Front and rear views.

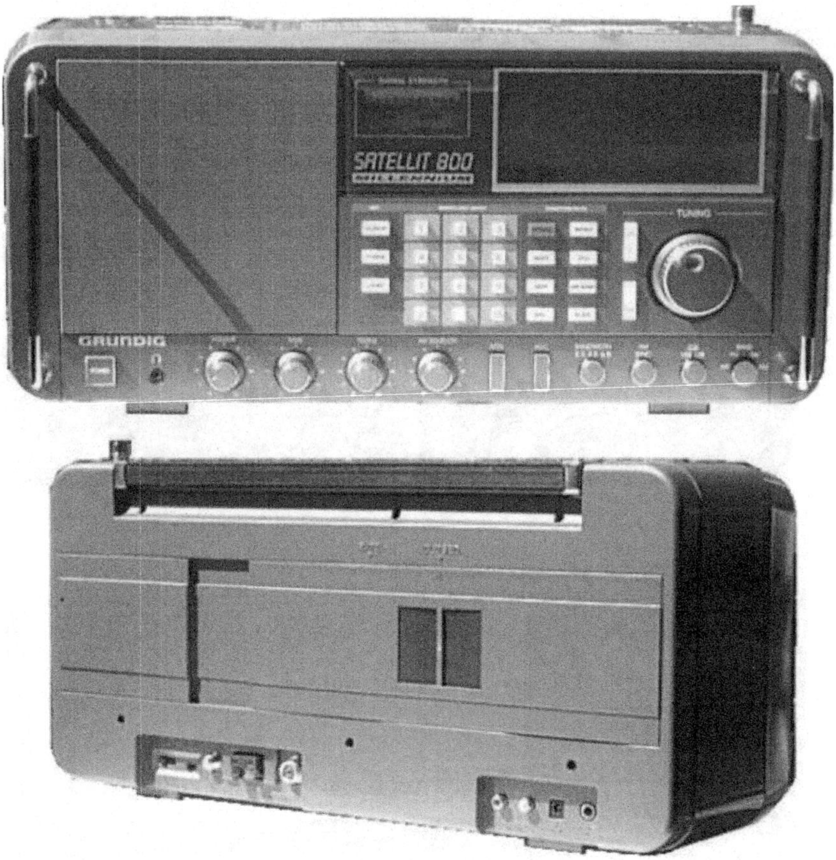

Satellit 800
Photo above: A look inside . . .
Photo below: packaging

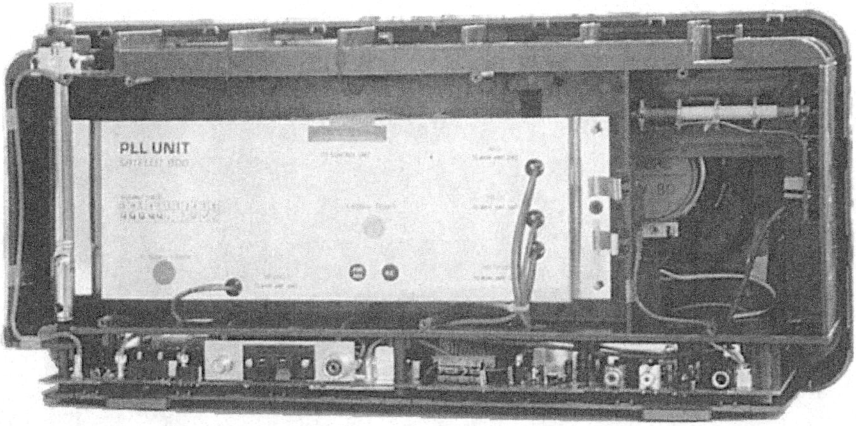

Satellit 800
Controversial imprint „Under License from ...“

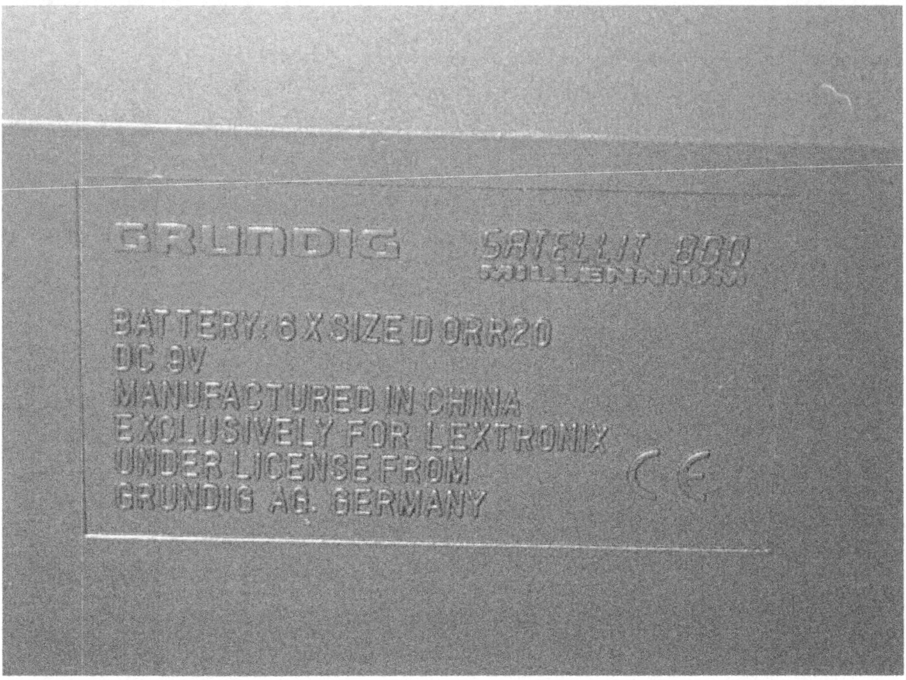

Model	IF Bandwidth at 6db	AF Power	AM IF (1.+ 2.)	Image Reject.
205	6,8kHz	2 Watt	460kHz	46db
208	4kHz / 6,8kHz	2,5 Watt	1,85MHz/460kHz	48db
210	4kHz / 6kHz	2,5 Watt	1,85MHz/460kHz	50db
1000	4kHz / 6kHz	4 Watt	1,85MHz/460kHz	50db
2000	3,4kHz / 6kHz	4 Watt	2MHz/460kHz	52db
2100	3,4kHz / 6kHz	4 Watt	2MHz/460kHz	52db
3000	3,4kHz / 6kHz	5 Watt	2MHz/460kHz	56db
3400	3,4kHz / 6kHz	5 Watt	2MHz/460kHz	56db
1400	4kHz	4 Watt	2MHz/460kHz	52db
2400	4kHz	2x4 Watt	2MHz/460kHz	52db
4000	4kHz	2x4 Watt	460kHz	50db
300	4kHz	2 Watt	460kHz	50db
400	4kHz	3 Watt	54,5MHz/450kHz	54db
600	2,7kHz / 4,8kHz / 6kHz	8 Watt	54,5MHz/450kHz	70db
650	2,7kHz / 4,8kHz / 6kHz	8 Watt	54,5MHz/450kHz	70db
500	3,5kHz / 6kHz	1,5 Watt	54,5MHz/450kHz	66db
700	3,5kHz / 6kHz	1,5 Watt	54,5MHz/450kHz	70db
900	3kHz / 5kHz / 6,8kHz	3 Watt	54,5MHz/455kHz	unknown
800	2,3kHz / 4kHz / 6 kHz	2 Watt	55,845MHz/455kHz	63db

Remarks:
Third bandwidth „superwide“ on model 3000, 3400, 600, 650, 800 and 900 is regarding to AF (Audio-part) only.

First AM IF at Satellit 210 Amateur is 2,5 MHz due to running the 160 meter band (± 1.8 MHz) as double conversion.

Audio output power listed above, indicates continues net power (regarding German Industry Standard), operated by AC.

All double conversion Satellit made until 1983 were single conversion radios below 5 MHz (5000 kHz) except 210 Amateur!

Newer Satellit (400,500,600,650,700,900) are using double conversion 1.6 MHz to 30 MHz

Satellit 205, 300 and 4000 were single conversion anyway.

Satellit Antennas

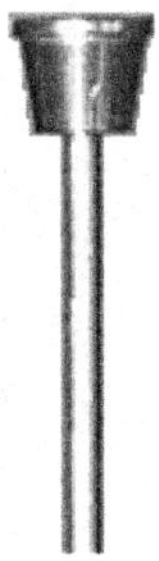

12 Elemente

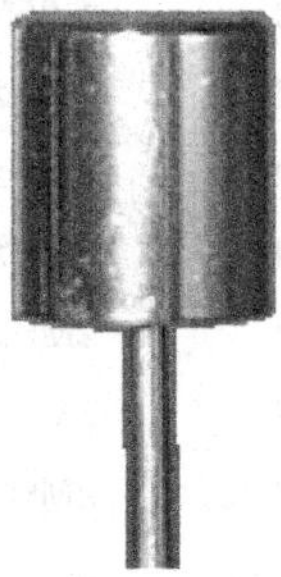

13 Elemente

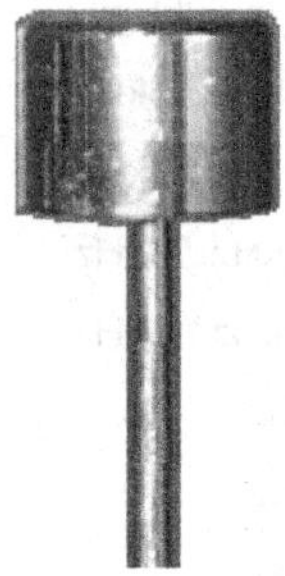

13 Elemente

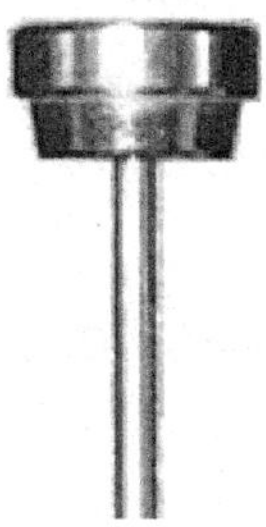

11 Elemente

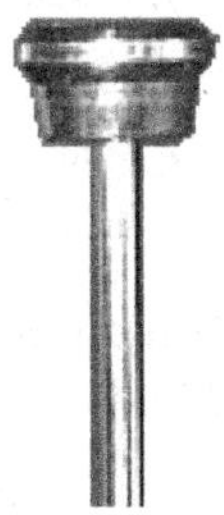

8 Elemente

Satellit Antennas

12 Element antenna: (taken from Ocean Boy 204)
Satellit 205

13 Element antenna with big tip
Satellit 208 / 210 / 1000 / 2000 / 2100 / 3000 / 3400

13 Element antenna with smaller tip
Satellit 1400 / 2400 / 600 / 650 und 800:

8 Element antenna
Satellit 300 / 4000

11 Element antenna
Satellit 400 / 500 / 700

9 Element antenne
Satellit 900

* last series Sat. 1400 and 2400 has been delivered with small antenna tip.

Mutants

205	German model, Satellit stated under the main scale
205 a	Two screws (instead one) to fix the hatch in the back
205	International model, Transistor 5000, different scale
205	International model, two screws, launched fall 1965
205 A	Radio-Amateur version, SSB build-in, other bands on drum
208	Body in grey, silver loudspeaker grill (rainbow) 1967 only
208	Body in black / silver (metal rainbow grill) national model
208	As above, only figures on the scale. International model
208	Body wooden-like
208	As above, only figures on the scale. International model
210	Body black / silver (metal rainbow grill)
210	As above, only figures on the scale. International model
210	Body wooden-like
210	As above, only figures on the scale. International model
210	Radio amateur version, HAM bands at the drum, black only
1000	Loudspeaker metal covered (mteal rainbow grill)
1000	As above, International model, called TR 6002. Int. model
1000	All in black (Profi-look), none aluminium-covered
1000	As above, called TR 6002. Figures only on scale. Int. model
1000	Loudspeaker wooden-like
1000	As above, called TR 6002, Figures only on scale. Int. model
2000	Loudspeaker metal covered (rainbow grill)
2000	Loudspeaker wooden-like
2000	Profi-Look, mainly in Black
2000	Loudspeaker cover in plastic made, edges in aluminium
2100	Loudspeaker cover in plastic made, edges in aluminium
2100	All in black (Profi-Look)
2100	Loudspeaker cover like Sat 3000
2100	As above but all knobs in silver (like Satellit 3000)
3000	Body in black, knobs silver
3000	Body copper colored ! Knobs also. Very last edition.
3400	Body in black, Loudspeaker grill coarse meshed
3400	Body in black, Loudspeaker grill fine meshed with silver ring
3400	Body in brown, Loudspeaker grill fine meshed, no ring!
1400	Body in black, Loudspeaker grill coarse meshed + silver ring
1400	Body in brown, Loudspeaker grill fine meshed + silver ring
2400	Body in black, Loudspeaker grill coarse meshed + silver ring
2400	Body in brown, Loudspeaker grill fine meshed + silver ring
2400	Body in
4000	One version known only
600	First 4500 units with gold scale, end 1983 made
600	Black scale starting with serial # 804501 (1984 to 1985)

Mutants

650	Blue-grey, Professional to 26.1 MHz, German model (1986)
650	Blue-grey, as above but International to 30 MHz, (1986)
650	Anthracite, Professional to 26.1MHz, German m. (fall 1988)
650	Anthracite, International, SW to 30 MHz (fall 1988)
650	Anthracite, replaces Prof. & International (fall 1990)
650	Anthracite, Final Edition , engraved edition number (1991)
300	Body in silver, all buttons in silver too.
300	Body in black, most buttons in grey
300	Model 300 a, better Filters (Ceramic)
400	Professional for native market to 26.1 MHz
400	International for export to 30 MHz
500	Professional, to 26.1 MHz (1988 to fall 1990)
500	International, to 30 MHz (1988 to fall 1990)
500	Italic, SW from 3950 to 26.100 kHz, LW to 302 kHz only
500	Late model, replaces Professional & International (fall 1990)
700	Normal model
700	Italic, SW from 3950 to 26.100 kHz, LW to 302 kHz only
900	One version known only (made in summer 1995)
800	First edition (end 1999)
800	Second & Third edition (fall 2000 & summer 2001)

Before you buy – check this

Good condition expected. Missing or broken antenna, damaged Loudspeaker-grill reduces value much. Also broken dial string, none working S-Meter or broken knobs / switchers.
Original papers, carrying bag or other accessories like SSB-unit, original box etc. increase collectors value dramatically !

Very good condition without any damage = 100%

Antenna damaged: - 20%
Wrong Satellitantenna: - 5%
Dial cord ripped: - 20%
Damaged body: - 15%
Loudspeaker grill with dents : -10% to -20%
Letters ribbed off: - 10%
S-Meter broken: - 10%
Contact problems: - 10%
Missing or wrong knobs or switchers: - 15%
Without AC powerpack: - 10%
Not all functions working: - 30%

40 % or more of deductions = Junk!

Owners manual included +10 %
All papers included: + 20%
With original carrying bag: + 20%
Other accessories: + 15%
Original box: + 20%

Example # 1: Satellit 210 with case and all papers is $500 worth. Just the radio, without any accessories, indeed with broken antenna at best$150.

Example # 2: Satellit 700 excellent condition = $250
Same radio with ribbed off letters and tiny dents in the loudspeaker grill = $250 - 10% (letters) – 10 % (dents) = $200 only !

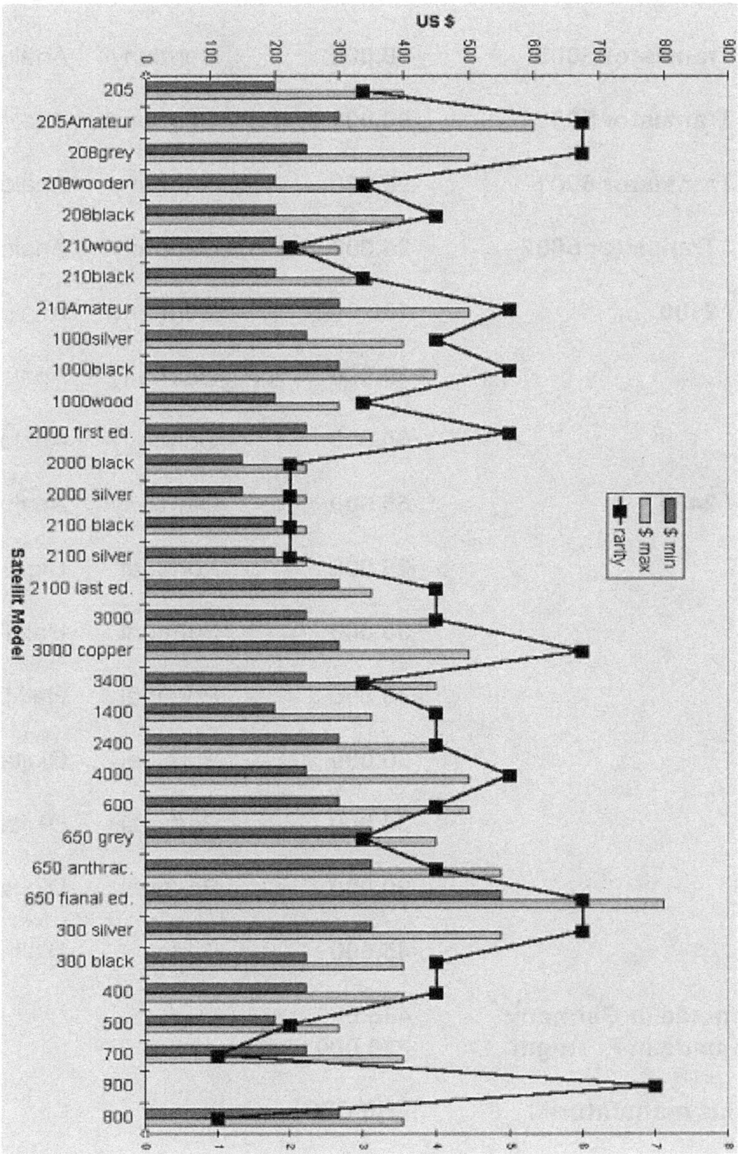
US $
Satellit Model
$ min
$ max
rarity
205
205Amateur
208grey
208wooden
208black
210wood
210black
210Amateur
1000silver
1000black
1000wood
2000 first ed.
2000 black
2000 silver
2100 black
2100 silver
2100 last ed.
3000
3000 copper
3400
1400
2400
4000
600
650 grey
650 anthrac.
650 fianal ed.
300 silver
300 black
400
500
700
900
800

Manufactured 1964 to 1996

Satellit model	manufactured	made in	technic
205 / Transistor 5000	50.000	Germany	Analog
208 / Transistor 6000	50.000	Germany	Analog
210 / Transistor 6001	90.000	Germany	Analog
1000 / Transistor 6002	25.000	Germany	Analog
2000 / 2100	100.000	Germany	Analog
3000	30.000	Germany	Analog
3400	65.000	Germany	Analog
1400 / 2400	55.000	Germany	Analog
4000	25.000	Portugal	Digital
600	30.000	Portugal	Half Digital
650	70.000	Portugal	Half Digital
300	30.000	Portugal	Digital
400	30.000	Portugal	Digital
500	60.000	Portugal	Digital
700	45.000	Portugal	Digital
Total made in Germany	465.000		
Total made in Portugal	290.000		
Satellits manufatured	755.000		

Chronical of Satellites:

1964	205	one screw			
1965	205 (a)	two screw			
1966	205 (a)				
1967	208	grey			
1968	208				
1969	210				
1970	210				
1971	210				
1972		1000			
1973		2000			
1974		2000			
1975		2000			
1976		2100			
1977		2100	3000		
1978		2100	3000		
1979		2100	3400	1400	2400
1980			3400	1400	2400
1981			3400	1400	2400
1982			3400	1400	2400
1983	300 silver		600	4000	
1984	300		600	4000	
1985	300 (a)		600	4000	
1986	400		650 grey-blue		
1987	400		650		
1988	400		650		
1989	400		650 anthracite		
1990	500		650		
1991	500		650		
1992	700		650 (final Edition)		
1993	700				
1994	700				
1995	700	900 (only ten prototypes)			
1996	700				
1997					
1998					
1999					
2000	800				
2001	800 (Third generation)				

Extra Pics

Young Lady from Fürth (Bavaria), Ingrid Finger won an international beauty contest. Grundig also recognize her ...

Prospectus 1967. Pic below: Case for Satellit 208/210

Extra Pics

This (dark blue colored) folder contains all documents regarding Satellit 650 Final Edition

Case for Satellit 2000/2100
Pic below: car adapter to run the first Satellits by 12 volts.

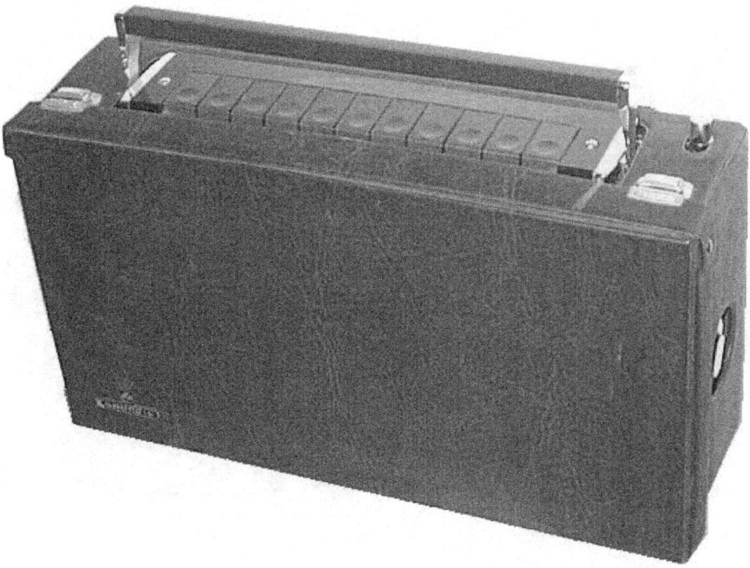

Extra Pics

Carrying bag (case) for Satellit 205 / Transistor 5000

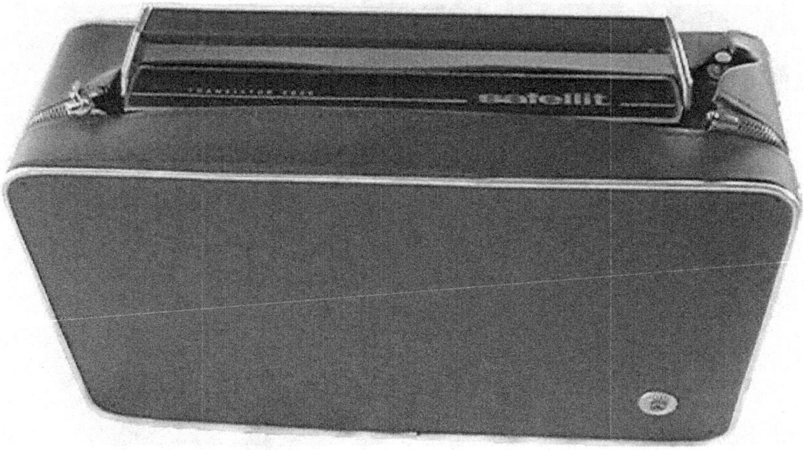

LC 24-10 EEPROMs (memo-chips) for Satellit 700

Pic above: Grundig facotry in the 1960s
Pic below: Advertising in the 1990s

After fitting or changing the batteries, arbitrary numbers appear.

FUNCT. Function selection (time/date) Entry of data

SELECT. Selecting the position, which has to be set

( ) Press the button as long (indication advances in ½ second rhythm), or as often, until the desired indication is reached.

(X) Press once

12 Indication blinks in ½ second rhythm

1) By actuating the "SELECT." button after the minutes have been set it is possible to start the clock **exactly to the second** (e. g. by means of radio or TV time check).

2) If the indication "time" is desired at once, then the "FUNCT." button must be pressed **4 times.** The appearing symbols have no function in this unit.

Attention: After adjustment has been completed, the "SELECT." button must be pressed repeatedly until no sign is blinking.

Setting the clock time

GB

FUNCT. S 1	SELECT. S 2	Indication	Explanation
—	—	10:25 08	Initial state: arbitrary time indication
—	(X)	10:25 10	Ready for setting the hours
( )	—	22:25 35	Setting hours, clock function not interrupted
—	(X)	22:25 40	Ready for setting the minutes, clock function not interrupted
( )	—	22:06 00	Setting minutes, clock function interrupted, seconds at zero
—	(X)	22:06 00	Setting process concluded, clock function started, indication "time". 1)

Setting the date (week-day, calendar day, month) (German week-day letters)

FUNCT. S 1	SELECT. S 2	Indication	Explanation
—	—	22:06 30	Initial state: "time"
(X)	—	MO 8 2	Indication: date
—	(X)	MO 8 2	Ready for setting the week-day
( )	—	DO 8 2	Setting week-day
—	(X)	DO 8 2	Ready for setting the calendar day
( )	—	DO 12 2	Setting calendar day
—	(X)	DO 12 2	Ready for setting the month
( )	—	DO 12 11	Setting month
—	(X)	DO 12 11	Setting concluded, indication "date". Changes after approx. 30 secs. back to indication "time". 2)

Extra Pics

The last 1000 Satellit 650 radios made, called Final Edition, each were delivered with this headphone.

Max Grundig

May 7, 1980 (Nürnberg) December 8, 1989 (Baden-Baden)

Extra Pics

In 1995 a new generation of portable shortwave radio has been developed by Grundig Yachtboy 500(below) and Satellit 900 (above).

Extra Pics

Grundig HQ located in city of Fürth, Kurgartenstr. 37
Pic below: At the 1st floor, Satellits has been made, at the 5th flook - they has been designed!

Extra Pics

In 1963 black (top), in 1986 grey-blue (below).

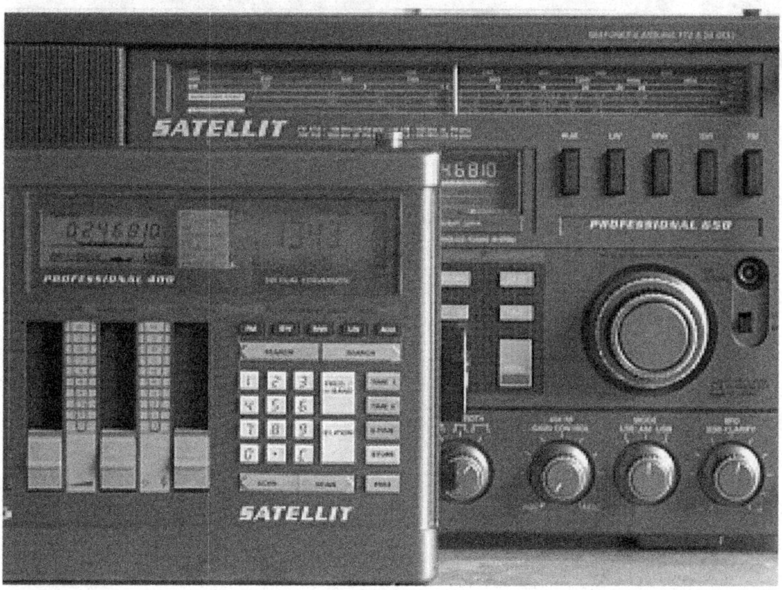

Extra Pics

Grundig advertising at Munich Central Station.

Advertising 1970

List of references

Grundig Revue, 1964 / 1965
Grundig, Technical info 1968 Notebook 1
Grundig, Technical info 1968 Notebook 4
Grundig, Technical info 1969 Notebook 1-2
Grundig, Technical info 1972 Notebook 2
Grundig, Technical info 1973 Notebook 3
Grundig, Technical info 1976 Notebook 3
Grundig, Technical info 1979 Notebook 5
Grundig, Technical info 1981 Notebook 3
Grundig, Technical info 1984 Notebook 3
Grundig, Short-wave guide, 1967
Grundig, Short-wave guide, 1969
Grundig, Short-wave guide, 1977
Grundig, Short-wave guide, 1979
Grundig, Short-wave guide, 1990
Grundig Owner's manual Satellit 208
Grundig Owner's manual Satellit 210
Grundig Owner's manual Satellit 2000
Grundig Owner's manual Satellit 2100
Grundig Owner's manual Satellit 3000
Grundig Owner's manual Satellit 3400
Grundig Owner's manual Satellit 1400
Grundig Owner's manual Satellit 600
Grundig Owner's manual Satellit 650
Grundig Owner's manual Satellit 4000
Grundig Owner's manual Satellit 300
Grundig Owner's manual Satellit 400
Grundig Owner's manual Satellit 500
Grundig Owner's manual Satellit 700
Grundig Owner's manual Satellit 900
Grundig Owner's manual TN 12
Grundig Owner's manual SSB unit
Grundig Catalogue Satellit 2100, 1977
Grundig Catalogue Satellit 3000, 1978
Grundig Catalogue Satellit 500, 1989
Grundig Catalogue Travel-super 1981 / 1982
Grundig Catalogue Radios and Recorder 1982
Grundig Catalogue Radios and Recorder 1983
Grundig Catalogue Radios and Recorder 1984
ADDX Kurier, October 1991
ADDX Kurier, March 1992
ADDX Kurier, October 1993
ADDX Kurier, September 1995
ADDX Kurier, October 1995
ADDX Kurier, November 1996
Rundfunk International 1990
Rundfunk International 1991
Radio Welt April 1992, Notebook 4
Radio handbook of wholesale trade VDRG e.V. 1965 / 1966

Useful Help

Types of Batteries
D-size, = Mono Battery
C-size, = Baby Battery
AA-size, = Mignon Battery

German DIN socket
Round, 5 pole standardised connection for audio equipment

Sine output power
German standardised norm for amplifier net output

RF = Radio frequency (receiving part of a radio)
IF = Intermediary frequency (filtering part)
AF = Audio frequency (audio amplifier part)

IFA = German most important fair for Radio and TV

DAB = Digital Audio Broadcast

DIN = German Industry Norm

cm = 2,54 cm equal to one Inch

mtr = 100 cm, equal to 3 Feet and 3 Inch

mtr Band = shortwave band
300.000 divided by frequency = mtr band !

All indicated "**new prices** " are stated in DM (German Mark) due to change of Dollar rate. In 1964 the rate was another as today or in the 1980s.

Satellit has not been translated to " Satellite ", because it's a registered trademark.

Made in the USA
Monee, IL
06 August 2026

59003505R00085